FIGHTING

CHURCH FENTON

The Story of

Royal Air Force Church Fenton (1937 - 1989)

by

Peter D. Mason

Pegasus Studio

First published 1989

Soft cover ISBN
0 9513645 0 2

Printed by A1 Press, Thorp Arch Trading Estate, Wetherby, West Yorkshire, LS23 7BJ

Fighting
Church Fenton

Spitfires of No. 19 Squadron on patrol
(Photo: Courtesy Imperial War Museum)

INTRODUCTION

It is a privilege for me to live in a village bearing the same name as Royal Air Force — Church Fenton, both located in a region steeped in military history and tradition. Consequently, it is not a difficult task to appreciate the long and distinguished record of this dual-role fighter-trainer base, now in excess of its 50th Anniversary. Although geographically remote from the southern counties and the critical Battle of Britain, the Station played a major role in the defence of the region, and the regrouping and re-equipping of depleted squadrons in the aftermath of the debacle following the Battle of France. Church Fenton also played host to the formation of several famous squadrons — No. 306 (Polish) . . . .No. 71 Eagles . . . . No. 242 (Canadian) . . . . and No. 249, to name a few, who later distinguished themselves in the "Battle" and other theatres of war.

The book narrates these epic times with personal first-hand experiences from notable personalities balanced with accounts from the trainee-newcomer.

It is very apparent as revealed from my many interviews, and world-wide correspondence as far afield as Australia, New Zealand, Canada and Poland, that Church Fenton has been a very efficient and happy Station — in peace and in war.

Many ex-Fenton personnel, from the mists of time, have made the sentimental journey back to base to relive experiences — indeed some have responded enthusiastically to the suggestion of a Reunion in the future.

In compiling this story I have tried to be factual, based on the personal account and Squadron Record — rather than imagery! The reader may ask of the motivation behind the composition of a historic narrative of what appears to come to be an ordinary station of the Royal Air Force in North Yorkshire.

There are two reasons attributable for the two years of dedicated hard work. An early pre-war indoctrination by an enthusiastic father who visited at every opportunity the Station Air-Days and the Alan Cobham Air Pageant at nearby Sherburn . . . the unforgettable spectacle of the German dirigible "Hindenburg" as its massive bulk made the slow, majestic flight over Leeds . . . .the Spitfires of No. 72 Squadron 'beating up' the Station hangers — a rich kaleidoscope of events.

A more recent explanation would be the chance remark made in the 'local' by a fellow villager — "You know, the region is steeped in military history and tradition which should be recorded whilst there are still folk around to tell the tale". This observation was further emphasised in conversation with the older contributors when plied for details — "Sorry, but the memory fades!"

Whatever the reasons for my unstinted enthusiasm for aviation, the fraternity has provided many years of pleasure and friendships, for which I owe a deep debt of gratitude. Although this epistle was never intended as a candidate for the 1989 Booker Prize, I sincerely hope that the reader derives as much pleasure from its pages as I enjoyed putting it together.

Peter D. Mason
Church Fenton — May 1988

ACKNOWLEDGEMENTS

The Editor wishes to thank all those who very kindly provided material, reminiscences, narratives, data and documentation in order to make this book possible; also for providing access to private correspondence and photographic albums, sometimes within the warmth and hospitality of their own homes and offices. It has been particularly gratifying to me that many contributors, of all ranks and denominations, have entrusted documents and personal photographs to my care without question or enquiry — all being of a rare and irreplaceable nature. So great has been the generosity and kindness of contributors that it is an impossible, and unfair task to single out any for special mention — excepting perhaps Mrs Muriel Nicolson for her wonderful home cooking!!

I would mention initially the various departments of the Royal Air Force. Group Captain C. J. Sturt OBE MBIM RAF (Officer Commanding Royal Air Force Church Fenton 1985 — 1987) for his very kind assistance, advice and cooperation, photographic help and access to Squadron Records. My thanks also to his successor, Group Captain D. A. Baron OBE RAF for his assistance.

Without the very helpful 'long-arm' of the Ministry of Defence Personnel Centre, Gloucester, who traced the former Station Commanders of RAF Church Fenton, the excellent response and their memoirs would not have emerged.

My thanks to Air Commodore H. A. Probert MBE MA., Air Historical Branch, MOD., London, for the provision of various AIR documents and advice. My thanks also to members of his staff for their kind cooperation.

To Wing Commander R. P. Beamont CBE DSO DFC DL FRAeS for his kind advice and for the provision of the FOREWORD and photographs. My sincere thanks also to Wing Commander P. B. Laddie Lucas CBE DSO DFC, author and publisher, for his very helpful advice and suggestions and Squadron Leader Jeremy Howard-Williams DFC for his cooperation and advice.

I am indebted to the Royal Australian Air Force Historical Section, Canberra, Australia, for permission to reproduce excerpts of Squadron Records for the chapter relating to No. 456 Royal Australian Air Force.

To the Polish Institute and Sikorski Museum my grateful thanks for their painstaking research on my behalf, and to the Officers of the Polish Air Force for generously contributing personal records and photographs. Also for kind assistance from the Polish Air Force Association, London and Leeds.

For the long interchange of correspondence and details I extend my grateful thanks to Mr Charles Sweeny, London, in regard to the chapter concerning the Eagles Squadron, also to Mr. Edwin D. Taylor of the Eagle Squadron Association, U.S.A., the American Embassy, London, and Mrs Vern Haugland for her cooperation with extracts from her husbands book 'Eagles Squadron'.

My sincere thanks to Squadron members for their fine personal reminiscences ... George Price ... Tim Woodman ... Darky Hanson ... Barry Radley ... Ron Hodgen ... Ken Holden ... and many others not mentioned — but by no means forgotten.

Mention must be made of contributors who are established authors in their own right who have kindly permitted extracts from their books, as follows:-
Marshall of the Royal Air Force Sir John Grandy, GCB GCVO KBE DSO - for notes on his tenure with No. 249 Squadron at Church Fenton as Squadron -Leader - as narrated in the book WINGS OF WAR by Wing Commander P.B. Lucas published by Hutchinson (1983)
Air Vice Marshal J. E. Johnson CB CBE DSO DFC (WING LEADER - Chatto & Windus 1956)

Group Captain Peter Townsend DSO DFC (DUEL IN THE DARK - Harrap 1985)
(DUEL OF EAGLES - Weidenfeld & Nicolson 1970)

Squadron Leader Jeremy Howard-Williams DFC (NIGHT INTRUDER - David & Charles 1976)

Mr Chaz Bowyer - (HISTORY OF THE RAF - Hamlyn 1977)
(FIGHTER PILOTS OF THE RAF - William Kimber 1984)
Mr J. J. Halley - (SQUADRONS OF THE RAF - Air Britain (Historians) - 1980)
Mr Simon W. Parry - (INTRUDERS OVER BRITAIN - Kristall Productions 1988)
Mr Brian Lunn - (AIRCRAFT DOWN - Air crashes around Wetherby 1939/45) - Hardwick Publications)
EAGLE SQUADRONS - (Vern Haugland - David & Charles Publishers)
UNDER THE WHITE ROSE - (by Frank Zeigler - McDonald & Co. - 1971)
TWENTY ONE SQUADRONS - (by Leslie Hunt-Garnstone Press - 1972)
BATTLE OVER BRITAIN - (by Frank K. Mason - McWhirter Twins)

PHOTOGRAPHIC ACKNOWLEDGEMENTS

I have pleasure in acknowledging the courtesy extended by the following sources for the photographic reproduction of their material:

The Editor - The Yorkshire Evening Press
The Editor - The Yorkshire Evening Post
The Editor - The Wakefield Express
The Keeper - Imperial War Museum
The Controller - Her Majesty's Stationery Office

I am indebted to Mrs Jean Holden, of Selby, for her many hours of high-grade typing on my behalf, and the loan of photographs for the chapters on No.616 Squadron and Group Captain Ken Holden DFC OBE.
My grateful thanks to the advertisers for their friendly and helpful contributions.

It is with sincere regret that I cannot include all the personal contributions presented by enthusiastic ex-Fenton personnel and others, and I extend my apologies to these very kind people for any omissions, which are controlled, unfortunately, by ever present economics and cost-related space. Perhaps a 'volume two' could be justified — food for thought!!

Editor - Church Fenton
January 1989

CONTENTS

- THE THREAT -

"As England - despite her hopeless military situation, still shows no sign of willingness to come to terms, I have decided to prepare, and if necessary to carry out, a landing operation against her. The aim of this operation is to eliminate the English Motherland as a base from which war against Germany can be continued, and if necessary to occupy the country completely"

ADOLPH HITLER - Directive No. 16 - 16 July 1940

- THE AWAKENING -

"Awake, arise, or be forever fall'n" ......
They heard, and were abashed, and up they sprung
Upon the wing, as when men wont to watch
On duty, sleeping found by whom they dread,
Rouse and bestir themselves ere well awake.

John Milton : Paradise Lost.

Air Chief Marshal, Sir Hugh Caswal Tremenheere Dowding, GCVO, KCB, CMB, Commander - in - Chief Fighter Command, 1940
(Photo: Courtesy Imperial War Museum)

FOREWORD

by

Wing Commander R. P. (Bee) Beamont, CBE, DSO + DFC + DL FRAeS

Pilot Officer R. P. Beamont (1940)
(Photo: Courtesy Wing Commander R. P. Beamont)

On 24 May 1940, the remnants of No. 87 Fighter Squadron began to gather at Church Fenton with, eventually, all their ground personnel, half their complement of pilots of only three weeks before - and no aircraft at all.

The squadron had been one of four permanently based in France in the Air Component of the British Expeditionary Force since October 1939, and with them had faced the onslaught of the Luftwaffe over France and Belgium which began on 10 May.

By 20 May, 87 we were down to only six fully serviceable Hurricanes but in those ten days they had fired their guns in more than eighty individual combats, and had lost, killed, wounded or captured - nine pilots.

It had been an impressive experience, but back in the well-ordered calm of Church Fenton, a classic permanent Royal Air Force station, there was a remarkable contrast.

The tone set by their Station Commander, Group Captain G. T. Richardson, and our own tremendous Squadron Commander, Squadron Leader 'Johnnie' Dewar, was simply - "well that was an interesting experience, but now you've all got your work cut out to re-train replacement pilots, work-up a full strength of new Hurricanes, and bring back the squadron to operational status within one month". With the Luftwaffe already making probing attacks over the Channel there was no doubt whatever in the mind of all squadron personnel where we would next be needed.

Meantime all ranks enjoyed Church Fenton, the pilots particularly frequenting 'The Old Ship' Inn at Ulleskelf, and only mildly resenting kindly reminders from the Station Commander that although we were all highly rated fighter pilots, roll-neck sweaters and top buttons undone were really not acceptable on his station!

Within a week the survivors from France were flying day patrols from Leconfield and night patrols from Church Fenton, in our brand new Hurricanes; new pilots were flying hard in formation and combat practice and by the end of June we were more than ready to return to the fight. 87 Squadron night operations, with Hurricanes based on Church Fenton, were probably the first of the war. We patrolled over Hull on a number of nights in June 1940 - I was unlucky not to get a shot at a Heinkel raider but he managed to ecape in cloud.

As mentioned earlier, our Commanding Officer was 'Johnnie' Dewar who, as a Wing Commander later in August, was the first officer of this rank to hold both the Distinguished Service Order and Distinguished Flying Cross in Fighter Command.

87 Squadron departed from Church Fenton on 5 July for Exeter, where it played a major role in the Battle of Britain, operating over Portland, Southampton and the South-West. We were only one of the many splendid fighter squadrons that were resident, and thus taken part in the history of Church Fenton, and we enjoyed the ambience of the station and the tremendous support of the Station Commander, and of the civilian community noted for its Yorkshire hospitality and friendliness.

In this book Peter Mason has shown, very successfully, and with dedicated research how Church Fenton, one of its finest stations has contributed to Royal Air Force history.

ROLAND BEAMONT
January 1988

Editor's Note
In the period 3.10.42 to 5.5.43 Wing Commander Beamont served as Commanding Officer (with the rank of Squadron-Leader) for No. 609 (West Riding of Yorkshire) squadron. This famous squadron is the subject of a separate chapter later in the book.

Squadron-Leader R. P. Beamont's personal Typhoon. (Photo: Courtesy 609 Collection)

1

INTERREGNUM

(Reproduced by permission of the Controller - Her Majesty's Stationery Office)

THE PRE-WAR PERIOD
INCEPTION & LOCAL MILITARY HISTORY

The year of 1933 saw the emergence of Hitler's Nazi Germany and the Third Reich, and the embryo stage of the German Luftwaffe under the supervision of (later) Reich Marshal Herman Göring. This was in contravention of the Versailles Treaty and under the clandestine guise of civil aircraft development.

The Royal Air Force since demobilisation in 1919 had existed on a 'shoe-string' and in consequence was both under-armed and ill-equipped. After heated debates in Parliament in July 1934, it was legislated, as a counter measure to German aeronautical activity, to form forty-one new squadrons - with facilities to keep them airborne. In 1935 plans carried forward and on 23 May the Government policy on Air Defence was published. A Treasury brief was given to the Air Ministry Directorate of Works to form a number of new "expansion works areas". Number 3 (Expansion) Works Area based at Lincoln was given the task of siting and contracting for airfields in Lincolnshire and Yorkshire - one of these being adjacent to the village of Church Fenton.

The intention to build here was not popular and generated considerable protest against the wastage of valuable agricultural land, particularly as an airfield already existed only 2 miles distant at Sherburn-in-Elmet. Such controversy about proposed aerodrome sites was operative throughout the country - but all protests were ignored. As late as 1937 however, senior Luftwaffe officers were invited to visit RAF Hornchurch - an important station in the London defences - to inspect Fighter Command, the prominent aircraft on display being the Gloster Gladiator and Hawker Fury fighters. These aircraft were slower than the serving Luftwaffe bombers and completely outclassed by the new Messerschmitt Bf 109 fighter, then in production. The Spitfire and Hurricane fighters for the RAF were not in evidence due to certain design problems on the engine. However, the first Hurricanes were delivered and operational a short time after the visit - a situation which appeared to be more than coincidence!!

LOCAL HISTORY

The area which was to be the environment of the new RAF Church Fenton was steeped in military history. The succeeding invasions of Britain had left their effects; Tadcaster and Cawood were originally Roman outposts instituted to guard river fords - the former over the Wharfe and the latter over the Ouse. They were appendices of the York garrison, to which they were connected by road.

The advent of aerial warfare brought new military activity to the area. During the 1914-18 war a Royal Flying Corps airfield was established at both Tadcaster and Sherburn-in-Elmet, the latter also providing facilities for the Royal Air Force. Now a further RAF station was to be instituted at Church Fenton. The station as planned under the original contract was completed in 1939 - although it had been in partial use for two years before this time.

1937 Whilst the construction of the aerodrome was still ongoing, the Station was officially opened on the 1 April, 1937 under the control of the Air Officer Commanding Fighter Command, initially in No. II Group and on 5 May to No. 12 Group. On 19 April the first Station Commander — Wing Commander W. E. Swann -assumed command. In the next two months took on the role of an operational flying station with the arrival of the first two squadrons - No. 72 from Tangmere on 1 June, commanded by Sqadron Leader E. J. L. Hope AFC equipped with Gladiators Mk. 2. In July, Gladiators Mk 2 of No. 213 squadron arrived from Northolt commanded by Squadron Leader J. H. Edwards-Jones. Both squadrons trained at the station for the rest of the year. Shortly after arriving, No. 213 won the Sassoon Trophy, a considerable acheivement for a newly formed squadron, thus establishing at an early stage the fine reputation of the station.

1938 This was to be the only peace-time year of operational training for Church Fenton prior to the Second World War, and was to prove one of interest and incident. Throughout the year the station functioned smoothly as a two-squadron unit, No. 213 leaving for Wittering on 18 May to be immediately replaced by No. 64 squadron, led by Squadron-Leader V. Beamish AFC, performing formation flying over the area in their Demon I aircraft on the day of their arrival. On 20 May the two resident squadrons combined to give an exhilarating air display to 3000 visitors who braved the rain, the first Open Day at Church Fenton — on the occasion of Empire Day. The aircraft performed air drill, synchronised aerobatics, attacks on a heavy bomber, and the dive-bombing and strafing of ground troops and transport.

Home defence exercises took place from 5—7 of August, No. 64 squadron were detached to Digby and No. 72 to Wittering. During the exercises No. 64 squadron lost four aircraft in unusual circumstances. On the night of the 7 August following take-off of four Demons, a dense fog descended and obscured a wide area of England. All the aircraft flew Southwards hoping to find clearer conditions before their petrol was expended. The effort was unrewarded, however, and with petrol supplies almost

Gloster Gauntlet
(Photo: Courtesy G/C Ken Holden)

Hawker Hart - Day Bomber
(Photo: Courtesy G/C Ken Holden)

exhausted, an order was received from the Sector Commander at Duxford, Wing Commander C.L. King MC DFC, to fly on a course of 225 degrees Magnetic — and abandon aircraft. Accordingly the four pilots and four rear gunners abandoned their aircraft and parachuted to safety. The four pilotless Demons flew on into the foggy night.

It was in the following month that early rumblings of approaching war were evident when the 'Munich Crisis' occurred. Acting on orders from No. 12 Group headquarters, all aircraft on station were armed and the base brought to a state of "immediate war readiness", which lasted from 24 September to 12 October.

Fortunately, another year was to elapse before the uneasy peace ended and the country was plunged into war, by which time the squadrons had been re-equipped with the Bristol Blenheim Mk. I twin-engined dual-role fighter-bomber.

Major changes in personnel took place in 1938. Wing Commander G.T. Richardson assumed command of the station from Wing Commander W. E. Swann, and the command of 72 Squadron changed twice from Squadron Leader E. J. L. Hope to Squadron Leader B. H. Rogers on 15 January to Squadron Leader R. B. Lees.

1939 January to 1 September — The Uneasy Peace

Without doubt, this period was the most progressive in the standard of new aircraft, No. 72 Squadron taking delivery of Spitfires in April — an aircraft to distinguish itself above all others in the ensuing momentous years. All the displays and the exercises performed were of special significance in respect to training. Both squadrons gave displays on Empire Air Days on 25 May at Church Fenton; a week earlier No. 64 Squadron commanded by Squadron Leader J. Heber-Percy, who had taken over from Squadron-Leader Beamish in January, led a massed flight of six squadrons over the Midlands. As in 1938, the squadrons were detached to other stations for the annual Home Defence exercises (11-12 August) No. 72 Squadron going to Wittering and No. 64 to Duxford. Church Fenton accommodated two Auxiliary squadrons — No. 602 from Abbotsinch

Royal Air Force Station, Church Fenton
EMPIRE AIR DAY — Sat,. 20 May, 1939

FLYING PROGRAMME — Part 1

Event No.	*Time*	*Event*
1	**14.00**	***Take-off by Wing of 18 Aircraft*** *The formation consists of 9 Blenheim aircraft of No. 64 (Fighter) Squadron and 9 Spitfire aircraft of No. 72 (Fighter) Squadron. After taking off these aircraft will form up into Wing formation (Squadrons line astern) and will return over the Aerodrome at 14-21 hours.* *NOTE: The Blenheim although originally designed as a Bomber is now, owing to its speed and manœuverability, also being used as a two seater fighter.*
2	**14.08**	***Individual Aerobatics — Gladiator Aircraft*** *The aircraft used in this event is a Gloster Gladiator single-seater fighter (Bristol Mercury Engine). The aerobatics performed will include loops, rolls, upward rolls and rocket loops.*
	14.10	***Visit of the Secretary of State for Air.*** *Arrival by air of the Secretary of State for Air, the Rt. Hon. Sir Kingsley Wood, and Chief of the Air Staff, Air Chief Marshal Sir Cyril Newall, GCB, CMG CBE AM*
3	**14.25**	***Demonstration of Speed by Spitfire Aircraft.*** *Three Spitfire aircraft will dive over the aerodrome to demonstrate the speed at which modern fighter aircraft now travel. Afterwards No. 72 (F) Squadron will land by Sections of three aircraft.*
4	**14.35**	***Air Drill by 4 Blenheim Aircraft of No. 64 (F) Squadron.*** *All the aircraft in this formation are fitted with wireless and orders will be transmitted from the ground to the pilots in the air. The aircraft will be seen to be changing formation.*
5	**14.55**	***Synchronised Aerobatics — Gladiator Aircraft.*** *Two Gladiator aircraft of No. 72 (F) Squadron will carry out aerobatics together. This event calls for very careful timing and the orders given by one pilot to the other will be broadcast over the loud speaker.*

6	15.15	***Air Race.***
		This will be a handicap race for different types of aircraft, the top speeds of which vary between 360mph and 120mph.
		The following types will take part in this event: Spitfire, Blenheim, Gladiaor, Battle, Demon, Hind, Magister and Tutor.
		The race will be flown over a triangular course, the aerodrome being one of the turning points.
7	16.03	***Fly Past by Formation of Battle Aircraft***
		These aircraft will arrive from another R.A.F Station and will fly over the Aerodrome in formation The Battle is a medium bomber type.
8	16.10	***Aerobatics by Tutor Aircraft.***
		The Tutor is a training aircraft fitted with an Armstrong Siddeley engine of 220 h.p.
9	16.20	***Demonstration by Lysander Aircraft.***
		The Lysander is used by army Co-operation Units and has the characteristics of quick take-off, slow landing speed and good control at slow flying speeds. This obtained with the help of slots and flaps fitted to the wings.
10	16.45	***Quick getaway by 3 Spitfire Aircraft.***
		This event represents the normal work of a Fighter Squadron in carrying out the interception of hostile bombers. On a given signal the pilots run to their aircraft, start the engines and take-off with the least possible delay. Once in the air, orders can be passed to the Fighters by radio-telephony.
11	17.00	***Attack by 3 fighters v. 1 Bomber.***
		One Blenheim of No. 64 (F) Squadron, representing a Bomber, will be attacked by 3 Spitfires of No. 72 (Fighter) Squadron.
12	17.15	***Fly Past of Aircraft in Park.***

and No. 609 from Yeadon, for their annual training, each squadron spending two weeks there during July and August. The last notable event, before general mobilisation was ordered on 1 September — was the movement of the station from the Command of 12 Group to that of 13 Group on 1 August. Invaluable work had been done at Church Fenton in the two years and five months since its opening, and the squadrons and ground personnel were well acquitted to perform the tasks about to be presented.

On 1 September, 1939, Hitler's armed forces invaded Poland, unprovoked and without warning, and the country underwent general mobilisation.

'WAR IS DECLARED AGAINST GERMANY'

So read the ominous entry in the stations Operations Record Book '540' on 3 September following Prime Minister Neville Chamberlain's 11.00am announcement to the nation. As an important Sector Station controlling Leconfield and the resident squadrons, Church Fenton was in a high state of preparedness, mobilisation being completed on 4 September 1939. On strength at this time were squadrons No. 72 and No. 64 and the 31 Anti-Aircraft Brigade Headquarters, with detachments of the Royal Tank Regiment and National Defence Corps as Army Guard. There were however problems with accommodation which was to remain throughout the war, despite the completion of several buildings and 'H'- type barrack blocks.

Camouflaging of the aircraft hangers and completion of two concrete runways, perimeter tracks and aircraft revetements were all declared serviceable by 2 November. No. 64 Squadron was called to investigate an enemy raid on 16 October thus gaining distinction of being the first operational squadron at Church Fenton. This call however proved to be abortive since it was outside the Sector, and no contact was made. The unit was later to leave for Evanton in Scotland with their Blenheim I aircraft for convoy patrol duties.

No. 72 Squadron also departed on 14 October for Sector station Leconfield and later Drem, near Edinburgh. It was at Leconfield whilst on defence operations that they had the honour of accounting for the first recorded 'kills' by a Church Fenton squadron, shooting down two Heinkel III's over the sea on 21 October.

Nearby Sherburn-in-Elmet was also under the control of Church Fenton during November, Leconfield being used as an advance base for No. 616 (South Yorkshire) Squadron (see chapter) in October, equipped with seven Gauntlets, four Battles and one Hind.

The first Canadian fighter squadron to be formed in the war was No. 242 in October, under Squadron Leader F. M. Gobeil armed with Battles, followed by Hurricanes Mk. I This squadron was destined to distinguish itself later in the Battle of Britain under its redoubtable Commander, Douglas Bader.

On 1 November the station was honoured by the visit of His Majesty, the late King George VI, who was impressed by a demonstration put on by aircraft of No. 64

Squadron. This was followed in December by a visit from the Air Officer Commanding in Chief of Fighter Command, Air Marsal Sir Hugh Dowding GCVO NCB CMG ADC., the culmination of an eventful and memorable year at the station.

1940 YEAR OF DESTINY

He that outlives this day, and sees old age,
Will yearly on the vigil feast his neighbours,
And say — Tomorrow is St. Crispian:
Then will he strip his sleeve, and show his scars,
And say — These wounds I had on Crispian's day.
Old men forget; yet all shall be forgot,
But he'll remember with advantages
What feats he did that day. Then shall our names
Be in their flowing cups freshly remembered.
We few, we happy few. . . .

This was to prove the most vital year in the history of Fighter Command with the occurrence of the Battle of Britain, and the victory achieved being the most momentous in British history. Although the geographical location precluded any direct participation in this action over London and the home counties, fighter squadrons that were formed, trained and equipped at Church Fenton played a leading role in the Battle. This included squadrons like No. 87 and 85 that had been resident here, depleted after the evacuation from France.

Resident squadrons here in early 1940 were; 242 — 64 and 72, the latter having recently returned from detachment at Drem. No. 72 Squadron — long resident here since 1937 moved on in March to Acklington. Shortly afterward, No. 64 also departed to Usworth (Newcastle) in May, followed by a further assignment to Kenley, where they played an important role in the Battle, now re-equipped with Spitfire I aircraft.

Although 242 had commenced to form in late 1939 they did not start serious training until January 1940 when they received their new Hurricanes I. It took them only twenty flying days after receiving their aircraft to become operational — this achievement being marked by a congratulatory signal from Air Officer Commanding 13 Group, Air Vice-Marshal R. E. Saul MC DFC.

By the end of the year, the squadron was to be credited with the destruction of over a hundred enemy aircraft.

New arrivals were Nos. 234 and 87 Squadrons (see Foreword by Wing Commander Beamont) in late May; 234 from Leconfield with their Spitfires, and 87 from the Battle of France arriving in fragmented fashion with their Hurricanes and ground crew, but having accounted well for themselves.

Two honours were bestowed to 87 in this period, when the Commanding Officer, Squadron Leader Dewar DFC was awarded the DSO on 31 May, followed on

8 June by a visit to Church Fenton by Marshal of the RAF, Lord Trenchard GCVO KCB CMG SO AFC when he congratulated the squadron for their fine work in France.

Marshall of the RAF, Lord Trenchard with G/C John Grandy (later MRAF John Grandy. (Photo: Courtesy 609 Collection)

The collapse of France brought another famous squadron to Church Fenton for rest and re-equipping. This was No. 73. Like squadrons 87 and 85 they had made a hasty withdrawal and evacuation but had lost 37 squadron members when their ship had been sunk by enemy action. They had also lost twelve pilots, killed or missing in France, and much of their equipment. However, they were soon operational again and ready to fly south to reinforce the already stretched air defences. Before leaving they were congratulated by members of Air Staff for their sterling service in France.

On 16 May 1940, the formation of No. 249 Squadron commenced under the command of Squadron Leader John Grandy (later Marshal of the RAF., Sir John Grandy GCVO GCB KBE DSO) — see separate chapter.

It is interesting to note the atmosphere prevailing at the station based on extracts from records in June and July of 1940.

++ "Operational activity to investigate of 21 unidentified raiders of whom five were intercepted, resulting in three combats with He. III.

++ On night of 18/19 June, enemy raiders dropped 15 high-explosive bombs and about 50 incendiaries near Stillingfleet — 3 miles west of Church Fenton. No serious damage reported.

++ A/C. 2 Wallace (a Canadian) flew off in a Blenheim at 00.50 hours — without permission, the aircraft landed with slight damage. It is believed that he flew the Blenheim for a wager, having had no previous flying experience.

++ Small calibre bombs dropped near Sherburn satellite station, Ulleskelf and Church Fenton railway station.

++ Frequent enemy patrols over coastal areas and reconnaisance over Church Fenton at night. Squadron No. 616 (South Yorkshire) from Leconfield shot down six raiders and three probables over the coast near Flamborough — all Ju. 88's.
No. 302 Squadron shot down a JU. 88 — the first to be destroyed by the Polish Air force. They later claimed two JU. 88 probables.

CHANGE OF ROLE.

On 10 August 1940 the station was transferred from 13 Group to 12 Group, the final transformation taking place in November, and the role changed from fighter defensive to the training of aircrew for night fighting, to be known as No. 54 Operational Training Unit. (This is dealt with at length in Chapter 8).

The location of the first night-fighter OTU at Church Fenton, plus the fact that the station was also the Sector Headquarters, resulted in visits from notable personages. These included His Royal Highness, The Duke of Kent (28.1.41), Her Royal Highness Air Commandant, The Duchess of Gloucester (11.2.41), the Air Officer Commanding-in-Chief Fighter Command, W. Sholto Douglas CB MC DFC (March 41— July 41 — 12 Dec. 41) Air Commodore Prince Bernhardt of the Netherlands (3.6.41), The Inspector General Air Chief Marshal Sir Ludlow Hewitt KCB CMG (23.1.42),the Air Officer Commanding 12 Group, Air Vice-Marshal R. E. Saul MC DFC, and on many occasions important visitors from Allied Air Forces.

The move of No. 54 OTU to Charter Hall on the 6 May 1942, brought to an end this eventful and extremely productive period of an operational fighter unit.

A Bristol Blenheim IV as used with No. 54 OTU (Photo: Courtesy British Aerial Museum)

THE OFFENSIVE ROLE (May 1942 — August 1945)

Enemy air activity was quite pronounced until October, 1942, the most severe raid taking place in Hull on the night of 19/20 May. Many minor incursions were made by German bombers, chiefly to the inland towns and cities, Leeds, York, Bradford and Doncaster being affected. As a consequence to these raids the vigilance of the defending squadrons had to be great, especially from 25 Squadron who were operating from the airfield, playing a major role in the Sectors defence, armed with Beaufighters, eventually being replaced by Mosquito Mk. II (F). This considerably increased their fighting efficiency.

Numerous night sorties were flown against enemy intruders (usually Dornier 217's) and these resulted in eight German aircraft claimed as destroyed and several others as probably destroyed or damaged. Four of their aircraft crashed with the loss of six crew, only one defence aircraft being lost in action.

There were three other squadrons stationed at Church Fenton in the latter half of 1942, these being:-

No. 488 (New Zealand) Squadron was formed on 25 June using Beaufighters Mk II (F) commanded by Wing Commander R. M Trousdale DFC.

No. 600 Squadron arrived 1 September with Beaufighters II (F), leaving on 14 November to make way for No. 183 Squadron. They commenced to form with Typhoons Ia and Ib led by Squadron Leader A. V. Gowers DFC, but due to initial difficulties with the type operational status was not achieved until 5 April 1943.

In August the station was honoured for the second time by the visit of Lord Trenchard, also Air Training Corps cadets visited the station for summer camps which boded well for RAF recruitment in the West Riding of Yorkshire.

1943 The early part of the year saw a further increase in the offensive role of the resident squadrons against the enemy, indicative of the changing aspect of the war. 'Intruder' and 'Ranger' patrols were intensified, commencing on 17 February by Mosquito's of No. 25 Squadron including No. 308 (Polish) Squadron with Spitfire V. b's led by Squadron Leader P. Niemiec. The number of intrusions quickly increased as the year progressed, the main Sector responsibility for defence now being vested with 604 Squadron, based at Patrington.

No. 26 Squadron arrived on 21 July with Mustang Mk. I aircraft led by Squadron Leader A. H. Baird, followed on 4 August by 96 Squadron (described in separate chapter).

The increased efficiency and success of the continental intruder raids by the Royal Air Force squadrons was beginning to be felt by the enemy.

Numerous trains, rolling stock, searchlights, airfields and factories were attacked and destroyed, or damaged. The effectiveness of these raids was countered in some measure by the continuing, if more sporadic, German bombing attacks on coastal towns and cities like Hull, Middlesbrough and Grimsby. The success of our defensive measures was diminished by the German use of our bombers returning from raids as a cover for their attacks, thus making interception very difficult for the British fighters.

Amid the concentrated flying activity it is not surprising that squadron losses were heavy. No. 25 Squadron lost nine aircraft during the year and No. 96 lost two in their short stay. Set against the losses of No. 25 was the notable record of intrusions successfully executed, and a total of eleven enemy aircraft destroyed, two probably destroyed and seven damaged in the eighteen month tour at Church Fenton.

On frequent occasions the weather interfered with the flying programme, much to the frustration of the grounded aircrew.

The cause of this was the infamous "Fenton fog", eloquently expressed by the chronicler of No. 25 Squadron — "mists breed in the low fields surrounding the airfield and steal softly in gently swirling masses across the camp. The south - west wind brings a reinforcement of industrial haze with a quick affinity for the moisture particles of the mist , bonding is rapidly complete and a blanket of obscurity muffles the sight".

Other important changes that took place on the station was the opening of No. 12 Group Defence School for training personnel in defence duties, lasting ten days duration, and in May Group Captain C. J. Collingwood DFC relinquished his command to Group Captain F. W. Stannard temporarily, when in June Group Captain G. A. L. Manton assumed full command.

The stations own "Wings for Victory Week" in May realised £10,096 — over £8000 more than the official target, which typified the spirit abroad at Church Fenton during this outstanding year.

1944 MAXIMUM OFFENSIVE

Again in 1944 the programme at Church Fenton was closely related to the progress of the war which had now reached a stage of maximum offensive, the zenith of which was the invasion of Normandy on "D-Day". Just as the station had played an important part in frustrating the German Operation "Sealion" in 1940, now the invasion process was reversed with Operation "Overlord" and the scene of the formation and training of some of the fighter units and squadrons that were to operate in the allied invasions and the consequent advance through Europe.

At the outset of the year preparations for "D-Day" were under way and the base commenced to operate a new establishment and organisation. In January No. 21 Base Defence Wing under the command of Group Captain W. G. Moseby DSO DFC, and No.141 Airfield commanded by Wing Commander G. O. Budd DFC started to form, and build up over the forthcoming months , and the arrival of mobile stations (CGI's, Signals and Operations Room) which were part of the composition of No. 21 Base Defence Wing, transferred on 3 March to 85 Group. This Group was in the allied Expeditionary Air Force, which was in turn controlled by No. 11 Group until after "D-Day", and was to be responsible for the defence of Allied positions in Europe during, and after, the invasion.

Several of the squadrons associated with No. 21 Base Defence Wing and No. 141 trained at Church Fenton for varying periods. No.'s 29, 409, 410 and 604 were here for brief visits and No. 264 Squadron operated here for five months from December with Mosquito XIII's, commanded by Wing Commander W. J. Alington DFC AFC., later Wing Commander E. S. Smith. Intensive training in preparation for the inevitable invasion were carried out, in conjunction with defence sweeps of the Sector, resulting in two kills — a Dornier 217 and a Heinkel 177. In June these units and associated squadrons provided cover for our forces throughout the Allied advance.

With the departure of the 85 Group units the station reverted to its previous status with an establishment for one night — fighter unit, and No. 307 (Polish) Squadron arrived to fulfil the commitment (refer to separate chapter). From May until September 307's primary role was that of intruder against Continental objectives. It is also worthy of note that No. 234 Squadron arrived early in the year armed with Spitfires to be deployed on intrusion and bomber support operations. And so another remarkable and successful year ended.

1945 PEACE IN SIGHT

The final months of the war brought about a natural decrease in the activities of the stations squadrons, and no further intrusion raids were carried out. The only operations were those in defence of the area. After a short but distinguished tour of duty at Church Fenton, No. 307 departed for Castle Camps. They were replaced by No. 456 (Australian) Squadron equipped with Mosquito XXX's and commanded by Wing Commander B. Howard DFC and remained here until 16 March, when No. 68 Squadron arrived with their Mosquito XXX's, to be disbanded a month later. Their place was taken by No. 125 Squadron — again with Mosquito XXX's. (The stories of both 456 and 125 Squadrons are related elsewhere in the book)

One other squadron was present at the station during this period — No.288 — equipped with Vengeances led by Squadron Leader Sampson, having previously been resident in 1943 and 1944 .

Squadrons 291 and 124 were based at Hutton Cranswick satellite station of Church Fenton, thus part of the Sector activities.

No. 125 Squadron was kept at a reduced state of readiness until the cessation of hostilities in Europe 'V-E Day' on 8 May, thus gaining the distinction of being one of the last squadrons maintained in defence of the country during the war. 'V-E' Day was celebrated with great enthusiasm at the station with a Dance, bonfire and Thanksgiving Service. With the war not yet over in the East however, the time of wholehearted rejoicing was delayed until 'V-J Day'. At this time there were 312 Officers and 1644 other ranks on strength, including many ex-POW's recently repatriated who were attending No. 112 Personnel Centre Refresher Course, which had started in June.

The feelings of the personnel on the station when total victory was announced is best summed up by an extract from the station Operations Record 540 — "15 August,

1945. At the close of the previous day, everything was peace and quiet but immediately the Prime Minister broadcast the news that Japan had accepted our surrender terms, the camp burst into life like a miracle to celebrate". The celebrations lasted two days and was a fitting climax to a great war effort in which thousands of personnel had participated at Church Fenton during the last six years.

THE POST-WAR PERIOD

Following the unconditional surrender of Japan came the inevitable criterion of change and adjustment to a peace-time role — a programme of training exercises.

The first 'At Home' was held on September 15 amid much previous Press publicity, and attended by 14000 visitors. The show was highlighted by a flying display given by No. 125 Squadron, and the final flypast given by most of the thirty different types of aircraft that had been on static display during the day. A final notable occasion in 1945 was the visit of Lord Montgomery who arrived in a silver Dakota on 26 October en route to receive the Freedom of the City of Huddersfield.

Visit of Lord Montgomery. October 1945
(Photograph and text commencing "The Post-War Period" are reproduced by permission of the Controller - Her Majesty's Stationery Office)

1946 Although the station had been functional for almost a decade, this year was the only the second full year of peace-time operation, and it was now that the pattern was set (until 1959) of two-squadron exercises and training basis. Undoubtedly the most memorable feature of the year was the arrival of jet-propelled aircraft in April — Meteor F3's of No. 263 Squadron, followed by No. 257 at the end of the year, having converted from Spitfires to Meteors. Church Fenton had the distinction of being one of the first Royal Air Force stations to house operational jet squadrons.

The station had many distinguished visitors during the year, including Marshal of the Royal Air Force Lord Tedder GCB DCL LLD

1947 In common with previous years, the interchange of squadrons continued and the arrival of No.19 Squadron on 29 April equipped with Hornets started an association between the squadron and the base that was to remain unbroken until June 1959. The severe weather conditions, heavy snowfall followed by extreme flooding of the area wreaked chaos of the flying programme in the first quarter of the year and it was not until the arctic conditions abated in April that the station became effective again. The now established 'At Home Day' in September was a great success, attracting over over ten thousand visitors. In October the station was re-organised on the 'Three-prong system' of Flying, Technical and Administration Wings, which operates to this day.

1948 Varied training was carried out by the squadrons with the greatest emphasis on cross country flights, gunnery practice and training with bombers. No. 41 Squadron flew to Fayid in October whilst No. 19 had a successful period of Annual Gunnery training at Acklington in May and June flying with B. 29's of the U.S.A.A.F. The largest single exercise the squadrons were involved in was Fighter Commands Annual Air Defence Exercise 'Dagger' in September.

1949 This again proved to be a full and varied year with Squadrons 19 and 41 carrying out a full training programme, basically as before, the Hornets being used additionally in two practice roles — as 'intruders' in March and 'night fighters' in August. The Mosquito Wing arrived from Coltishall in November, consisting of Nos. 23, 141 and 264 due to improvements being carried out there. Thus with five squadrons operating on the station at the close of the year a very busy state of affairs existed. In contrast to these added responsibilities, Church Fenton suffered a loss which changed the face of the camp. In March it was decided to post all W.A.A.F. personnel to Linton-on-Ouse with immediate effect. Some consolation perhaps was given to the now completely male personnel when Miss New Zealand presented the prizes at the Station Sports Day.

1950 With the continued presence of the Mosquito Wing until September, the arrival of No. 609 (Auxiliary) Squadron in October, and the fulfilment of a full training programme by 19 and 41 Squadrons, culminated in the highest number of flying hours from the base since the war. In the annual Air Defence Exercise

'Emperor', 19 and 41 Squadrons operated in the new role of low level intruders attacking targets such as airfields, ships and GCI installations from "nought" feet. During the weekends of Exercise Emperor No.'s. 609 and 607 Auxiliary Squadrons operated in the defence of the station. It was at this time that 609 was commanded by Squadron Leader A. Hudson DFC and equipped with Spitfires XVI's when they moved to Church Fenton, since the runway at Yeadon was too short to cater for the Vampire V's expected in November.

1951 This year was chiefly memorable due to the complete conversion of all the resident squadrons to jet-engined aircraft, Meteors Mk IV's or Mk. VIII's. No. 41 moved on to Biggin Hill in March, 609 by this time had eleven fully converted pilots to fly the squadrons new Meteor IV's. Involvement in the annual Air Defence Exercise 'Pinnacle' in September/October achieved a high degree of efficiency after some initial problems were resolved. Other notable events achieved this year were the fly-pasts over Leeds in July by 609 to aid the Territorial Recruitment Drive. The station was honoured by a visit on 6 June by the Fighter Command Air Marshal, Sir Basil Embry KBE CB DSO DFC AFC. The main runway 17—35, which had been the subject of concern for some time, was now scheduled for re-surfacing. Whilst work was being carried out, runway 07—25 was put to full use for training.

1952 Intensive training by the resident squadrons 609 and 41 continued with bomber affiliation with RAF and USAAF aircraft, high altitude intercep-tions, "rat and terrier exercises", controlled descents through clouds and gunnery practice. Air gunnery at towed targets using Meteors started in May.

Gloster Meteors formate over Church Fenton
(Photo: Courtesy 609 collection)

1953 The consequence of the arrival of No. 72 Squadron from North Wield with their Meteor VIII's greatly increased the station establishment in May. It will be remembered that this squadron was the first to operate from the base — with Gladiators in 1937, and were no strangers to Church Fenton. In May, work was started on the extention to runway 07—25. No. 19 Squadron excelled in the gunnery field, and received a congratulatory signal from the Air Officer Commanding No. 12 Group for the "top twelve shots", culminating in the award of the Dacre Trophy for efficiency for air firing.

609 had a good year, but narrowly missed winning the Esher Trophy for the most efficient Auxiliary squadron — losing to 615 Squadron.

In February Air Vice Marshal R. L. R. (Batchy) Atcherley, Air Officer Commanding No. 12 Group made his annual inspection, indicating satisfaction with the general progress of the station. He again visited the base on the 21 June to present the new squadron crest. The resident squadrons, in the absence of an 'At Home Day' at Church Fenton this year, provided aerial displays at 'away' bases of Duxford, Odiham, Hemswell, Scampton, Manby, Binbrook and Newton, to celebrate the commemoration of the Battle of Britain and the Coronation. Unfortunately over the year No. 19 Squadron suffered four fatalities in collisions and crashes.

1954 This was a very good year for 609 who deservedly were awarded the Esher Trophy. They consistently achieved a high number of flying hours (251 on the weekends in May alone) and an excellent standard of efficiency in training. The runway 07—25 was largely completed, and plans were drawn up for a Radio workshops, an Equipment Section extension, a new Airmens Mess, and additional Airmen's married quarters.

1955 The very high standards achieved by the squadrons over the past two years were maintained in 1955, a large degree of success being credited to the Wing in its role of 'interceptor', No. 72 excelling in the highest squadron and individual scores yet recorded for air-to-air firing. All three resident squadrons paid visits abroad — 609 for their Summer camp to Celle and Nos. 19 and 72 to Chievres, Belgium. A successful Fighter Command versus the Northern (Army) Command Boxing Tournament took place in August, attended by several local and Service dignitaries, including Bruce Woodcock, the former British Heavyweight Champion. In June the station passed from No. 12 into No. 13 Group, and changes in command occurred in each squadron. It was announced in October that No. 19 Squadron was to be re-equipped with Hunters F6, and 72 Squadron with Meteor NF 12/14. Both squadrons were to revert to an establishment of sixteen aircraft and twenty-one pilots. The conversion commenced in November.

1956 The gradual conversion was the dominating feature of the year, No. 19 proved slow due to shortage of aircraft, lack of spares, and atrocious weather, but happily were almost operational by the year end. Highlights were 609's summer camp at Malta and a short exchange visit between No. 19 Squadron and No. 350 Squadron Belgian Air Force, which took No. 19 to Beauvechain and the annual Air Defence Exercise "Stronghold" in September. Two ceremonies worthy of recall are firstly the visit of Marshal of the RAF, Lord Trenchard, on the 21 February, for a Memorial Service. The other was the presentation of the Standard to No. 19 Squadron on attaining its 21st birthday. Air Chief Marshal, Sir Donald Hardman KCB OBE DFC, a member of the squadron in 1918, made the presentation. He was supported by Air Marshal H. L. Patch CB CBE, the Air Officer Commanding in Chief Fighter Command and other dignitaries. After the ceremony which was attended by many ex-squadron members, Air Marshal Hardman endeared himself to all present by a demonstration of competence on the 'crash-box' of the Squadron taxi on departure.

1957 The year opened on a sad note at the station for notice was received that all Auxiliary squadrons were to be disbanded by 10 March. As a result No. 609 Squadron ceased flying on 11 January (see relevant chapter). As compensation for the loss of 609 a former resident squadron arrived — No. 85 — commanded by Wing Commander W. F. Binks DFC with Meteor NF 12/14's. Intensive training by both No. 19 and 72 squadrons continued throughout the year, highlighted by the annual Air Defence exercise 'Vigilant' from 24—27 May, during which a large measure of success was achieved despite the frequent out-performancing of the No. 72 Squadron Meteors. The exercise was witnessed by several important visitors including General Kramberg, the Commander in Chief of Fighter Defence Danish Air Force, and four American Colonels, two from SHAPE and two from Washington.

1958 Major exercises took place in June when the station squadrons participated in a NATO Exercise 'Full Play' and in sixty sorties the Wing claimed thirty four Canberras's and eight 'V' bombers as "destroyed". Because of the Middle East situation the Wing was brought to a state of one hours readiness on the 18 July, which continued until 6 August. Air Chief Marshal, Sir Thomas Pike KCB CBE DFC, Chief of Fighter Command and the Air Officer Commanding No.12 Group, Air Vice Marshal H. P. Fraser CB CBE AFC visited the station on 7 November to present the Ingpen Trophy for 1958 to No. 85. This trophy was an award competed for in the all-weather role with marks for cine, air firing and aircraft turnaround. However it was announced shortly afterward that 85 was to be disbanded and No. 89 at Stradishall were to be re-designated No. 85 on the 1 December. The squadron diarist wrote — "on the first day of the month the squadron won the Ingpen Trophy, and on the last day became non-operational — sic transit gloria".
The 'At Home' Day' on 20 September broke all records by attracting an attendance of twenty five thousand visitors, and realising £1000 in proceeds for the RAF Benevolent Fund.

1959 Since the station was commissioned in 1937 Church Fenton had been controlled by Fighter Command and had been one of the leading dual-role fighter/trainer bases during this period. This long association was finally severed in

July 1959 when squadrons 19 and 72 moved on to Leconfield, and Church Fenton became a unit of Flying Training Command. So ended an era which the station could be justifiably proud. The station now became the home of several important units, the principal ones being No. 23 Group Headquarters... No. 60 Maintenance Unit ... No. 2 Police District and Leeds University Air Squadron.

1962 - 1966 Still within Flying Training Command, the station became the home of No.7 Flying Training School equipped with Jet Provost aircraft under No. 23 Group Headquarters. During this period over 400 trainee pilots graduated and were presented with their'wings', among them many from the Commonwealth and other countries . For a brief period in 1966 an advanced training unit operated alongside No. 7 FTS using Vampire T. 11 aircraft.

In April 1966 Leeds University Air Squadron returned to the station, together with No. 9 Air Experience Flight. Both units used Chipmunk aircraft.
September 1966 saw the arrival of No. 6 Recruiting Region together with associated Schools Liaison Officers.

1967 - 1968 In January 1967 the Primary Flying Squadron arrived from South Cerney, and on 1 January 1967 the station was transferred from No. 23 Group to No. 25 Group. On 1 January 1968 the station was transferred once more to No. 22 Group and Training Command.

1969 -1973 The Aircrew Officer Training School moved from RAF Church Fenton to RAF Henlow where it merged with the Officer Cadet Training Unit, the transfer being phased over a period of three months. Church Fenton continued as a flying training school from September 1969 units on the Station included the Yorkshire Universities Air Squadron and No. 9 Air Experience Flight. The Helicopter Specialist Air Squadron was later renamed the Royal Navy Elementary Flying Training Squadron. Bulldog aircraft were introduced on the station in July 1973.

Squadron Leader L. K. Grybowski (ex. No. 306 (Polish) Squadron) congratulates his son John on graduation at Church Fenton 22.3.69 (Photo: by kind permission of the Yorkshire Post)

1974 - 1978 No. 2 FTS disbanded in November 1974 and shortly afterward the RN EFTS moved to RAF Leeming. On 21 March 1975 RAF Church Fenton closed, and for the next four years the airfield remained in use as a relief landing ground, under the control of Linton-on-Ouse. (See also Air Commodore J. A. Bell's account).

1978 STATION REACTIVATION In 1978, with the requirement formore pilots, additional flying schools were needed, the station was formally re-opened as No. 7 FTS on 2 April 1979 equipped with Jet Provost aircraft. The Jet Provost had made its debut at Church Fenton with No. 7 FTS in 1962 and had given outstanding service in the role of a basic trainer aircraft of the RAF since this time.

THE FUTURE In 1987 the Ministry of Defence selected the new Shorts Tucano as a replacement for the now obsolescent Jet Provost Mk 3 and Mk 5A. The Tucano will shortly be assuming the role of basic trainer over the skies of Church Fenton with its improved overall performance and efficiency, with low life-cycle costs combining the forgiving flying properties with superb aerobatic capability.

Other changes at Church Fenton are the take-over of the Supply and Engineering services by civilian contractors. This new period should present further opportunities and stability, and continued security for the nation in the provision of a high quality basic raining for the Royal Air Force for years to come.

The Shorts Tucano (for the technically minded)

++++ The power unit is a Garrett TPE 331-12B turboprop engine combining high performance and efficiency with low cost maintenance. Of 1100 SHP, it drives a four bladed Hartzell fully-feathering and reversing propeller of 2.39 metres diameter.

++++ Performance 270 knots - ISA at sea level. Wing span 37 feet, length 32.35 feet.

The Shorts Tucano — the new basic trainer of the Royal Air Force
(Photo: Courtesy Short Brothers Ltd. Belfast)

The Esher Trophy

.THE LORD ESHER TROPHY was presented in 1926 by Viscount Esher for annual competition between squadrons of the Royal Auxiliary Air Force. The trophy is a bronze figure, Perseus, standing on a base plate supported by pillars.
Since 1946 the trophy had been competed for annually on a basis of all-round efficiency, and awarded to the Auxiliary Fighter Squadron judged to be most efficient in all-round fighter squadron duties.

The Esher Trophy
(Photo: Courtesy 609 Collection)

HISTORY OF NO. 7 TRAINING SCHOOL

No. 7 Flying Training School was first formed at Peterborough on 2 December 1935. The role of the school was to train pilots and the aircraft used were the Avro Tutor, Hawker Hart, Hawker Audax and Hawker Fury. The course lasted 6 months and included weapons training, formation flying and tactics. The students gained their wings after successful completion of the first 3 months of the course and spent the last 3 months practising applied flying. The first course to go through the school consisted of 62 officers and 34 airmen pilots.

The school continued to train pilots for the Royal Air Force until May 1939, when it began what was to be a long association with the Royal Navy by taking on the new role of training pilots for the Fleet Air Arm. With the outbreak of war the length of the course was reduced to 4 months.

In January 1940 the name of the unit was changed to No. 7 (Service) Flying Training School and the Fairey Battle was added to its aircraft. Shortly after this in August 1940, the unit embarked for Canada as part of the expansion of the Empire Trianing Scheme. Whilst the unit was still en-route for its new home the name was again changed to No. 31 (S) Flying Training School and on completion of the move in January 1941 No. 7 Flying Training School was completely disbanded.

The school was refounded, again at Peterborough, in June 1942 with the title of No. 7 (Pilot) Advanced Flying Unit and was given the role of refreshing and acclimatising pilots trained in overseas schools. This unit was disbanded towards the end of 1944 but stayed at Peterborough to form the nucleus of No.7 (S) Flying Training School. Most of the students going through the school at this time were French, Dutch or Belgians, and the aircraft used by the unit were the Miles Master, Airspeed Oxford and the Supermarine Spitfire.

In early April 1946 the unit moved to Kirton-in-Lindsey and in the following year, after the departure of the last of the students of the Belgian Air Force, renewed its previous association with the Royal Navy. A naval course of 11 officers and 13 ratings arrived in November 1947 and trained in company with Royal Air Force students. The unit regained its original name of No.7 Flying Training School in 1948 when it moved to Cottesmore. Towards the end of 1952 the school was made responsible for the training of Royal Air Force students according to the "all through" sequence of training; the aim of which was to provide all stages of flying training up to "wings" standard at one school. The aircraft used to carry out this training were the Prentice, Harvard and later the Balliol. The responsibility for training naval courses passed to No.1 Flying Training School at Syerston.

The unit badge was approved by his Majesty King George VI in April 1949. It consists of a sprig of maple, a key, ward downward, and the motto: "Arming to Strike". The key, which is adopted from the armorial bearings of the Soke of Peterborough, signifies knowledge, and the maple commemorates the unit's residence in Canada during World War II

No. 7 Flying Training School was again disbanded in March 1954 as the result of the contraction of Flying Training Command. However, it arose Phoenix-like from the ashes for No. 202 Advanced Flying School which had been disbanded at Royal Air Force Valley. For the third time the school assumed the familiar role of training pilots for the Royal Navy. It was equipped with the Vampire TII and continued with the training already begun at Syerston on Provosts to "wings" standard.

Because of the ever decreasing number of Royal Naval students required for pilot training the Navy decided to carry out the whole Provost/Vampire sequence on one station. As a result of this all the naval personnel departed with their equipment for Linton-on-Ouse in January 1958. In June of the same year the remnant of No. 7 Flying Training School was augmented by the arrival of the personnel of No. 4 Flying Training School which had been disbanded at Worksop when that station closed down. The school continued the training of those pilot courses which No. 4 Flying Training School brought with them from Worksop and took on the subsidiary task of providing advanced flying for post-graduate studentsunder the title of No. 4/7 Flying Training School. In July 1960, the No. 7 Flying Training Sxhool element was officially disbanded.

In April 1962 No. 7 Flying Training School reformed at Royal Air Force Church Fenton with the responsibility for basic training of Royal Air Force pilots on the new "all through" Jet Syllabus, using Jet Provost Trainers Marks 3 and 4. No. 7 FTS disbanded yet again in November 1966.

On 2 April, 1979, No. 7 Flying Training School was once more reformed at Royal Air Force Church Fenton, flying Jet Provost Mark 3A and Mark 5A aircraft. The course at present lasts a total of 160 hours.

In 1987 it was decided to make the new Short Brothers Tucano the standard Royal Air Force trainer to replace the long-serving, but now obsolescent Jet Provost, which should begin to arrive towards the end of July, 1989.

ROLL OF STATION COMMANDERS
R.A.F. CHURCH FENTON

Rank	Name	Dates in Post	Present Rank/ Rank on Retirement
+ Wg Cdr	W E Swann	April 37 - 38	Wg Cdr
Wg Cdr	G T Richardson	38 - Jun 40	Deceased
Gp Capt	C F Horsley MC	June 40 - June 41	Deceased
Gp Capt	L G Nixon	June 41 - Feb 42	Deceased
+ Gp Capt	C J Collingwood (OBE) DFC	Feb 42 - May 43	Gp Capt
+ Gp Capt	F W Stannard	May 43 - June 43	Gp Capt
+ Gp Capt	G A L Manton	June 43 - Feb 44	Gp Capt
Gp Capt	B S Thynne AFC	Feb 44 - Jan 45	Deceased
Gp Capt	J G Hawtrey	Jan 45 - Mar 45	Deceased
+ Gp Capt	G C Tomlinson (OBE DFC)	Mar 45 - June 45	Gp Capt
Gp Capt	E W Whitley DSO DFC	June 45 - Nov 45	Deceased
+ Wg Cdr	E Plumtree (CB) OBE DFC	Nov 45 - Feb 46	Air Vice Marshal
Gp Capt	E W Whitley DSO DFC	Feb 46 - March 47	Deceased
Wg Cdr	G H Dyson MBE DFC	March 47 - Sep 47	Deceased
Wg Cdr	M D Thunder	Sep 47 - Sep 49	Deceased
+ Wg Cdr	E N Ryder (CBE) DFC	Sep 49 - Jan 52	Gp Capt
+ Wg Cdr	I S Smith (CBE) OBE DFC*	Jan 52 - July 53	Gp Capt
Gp Capt	J C Wells DFC	Jul 53 - March 54	Deceased
+ Gp Capt	C F Gray DSO DFC**	Mar 54 - Mar 56	Gp Capt
Gp Capt	L M Laws	Mar 56 - Mar 58	Deceased
+ Gp Capt	G D Sise DSO* DFC*	Mar 58 - July 59	Gp Capt
+ Wg Cdr	K R Bowhill OBE BA	Jul 59 - Nov 59	Wg Cdr
Wg Cdr	R. H. Blackmore	Nov 59 - Jan 61	Deceased
+ Wg Cdr	J E J Sing DFC	Jan 61 - Mar 62	Wg Cdr
Gp Capt	D R Stubbs DSO OBE DFC	Mar 62 - Sep 64	Deceased
+ Gp Capt	C C M Baker OBE	Sept 64 - Dec 66	Air Cdre
+ Wg Cdr	C F Green AFC	Dec 66 - Feb 67	Wg Cdr
+ Gp Capt	G J South DSO DFC	Feb 67 - Jun 69	Gp Capt
+ Gp Capt	J W Foster DFC AFC	Jun 69 - Jul 71	Gp Capt
+ Gp Capt	W G Abel	Jul 71 - Aug 73	Gp Capt
+ Gp Capt	D G F Palmer OBE	Aug 73 - Jan 75	Gp Capt
+ Wg Cdr	P N Farley	Jan 75 - Mar 75	Wg Cdr
Unit on Car and Maintenance		Mar 75 - Apr 79	
< Gp Capt	J A Bell OBE	April 79 - Dec 80	Air Cdre
+ Gp Capt	K R Briggs	Jan 81 - Jun 83	Gp Capt
< Gp Capt	R Humphrey ADC	June 83 - July 85	Gp Capt
< Gp Capt	C J Sturt OBE	July 85 - (Aug 87)	Gp Capt
Gp Capt	D A Baron OBE	Aug 87 - Present	Gp Capt

+ = appear in 1986 edition of RAF Retired List

< = Still serving RAF Officers

Decorations in brackets awarded after service at RAF Church Fenton.

2a

STATION COMMANDERS' MEMOIRS

GROUP CAPTAIN C.T. COLLINGWOOD, O.B.E., D.F.C.

(February 1942 — May 1943)

"I commanded this Fighter Sector Station in 1942 — 1943 which had a dual day-and-night fighter role, of which the night fighter role element was the principle activity. It was a long time ago, and as I never kept a diary, it is not easy to remember the squadron members serving on the station.

When I first took over the command, the day fighters were Hurricanes, the night fighter squadrons were Beaufighters, later to be replaced by Mosquitos. Towards the end of my tour of duty we were sent a squadron of the new Hawker Typhoon Grand Attack fighters, in training for the Normandy landings and subsequent advance into Europe. At this time we were equipped with Mosquitos which, very occassionally broke up in the air, normally with fatal results to the crew. Church Fenton lost the Squadron Commander and his air gunner in one of three unexplained crashes one night, and shortly afterwards a pilot in a day training flight reported that he was having difficulty in controlling his air-craft. He was ordered to fly over the station again for his passenger to eject by parachute and then land his aircraft, which was accomplished successfully. On examination, it was discovered that the aircraft undercarraige had lowered in flight and the port side door to the nacelle had been torn off in the slipstream. It had then been swept back and cut through the front spar of the tail plane, leaving it flapping about. The accident investigation branch was asked to inspect the damage. After checking up on the detailed reports on proving accidents, they came to the conclusion that the severing of both front and rear spars of the tail plane, had been the cause of the unexplained crashes. As it transpired, a modified strengthening of the locking mechanism of the undercarriage had already been initiated, so Church Fenton did not create a great revolution in Mosquito safety, but the mystery of the crashes was solved."

C.T. Collingwood
October 1987

GROUP CAPTAIN G. A. L. MANTON

(June 1943 — February 1944)
(now living in Australia)

"My period of command at Church Fenton was really one of training and consolidation. I well remember the night raids on the east coast — and Hull in particular — but No. 25 Squadron were not too successful as the A.I. was not very effective. As Station and Sector commander, life was very full. Hutton Cranswick as a Sector Station, took a slice of my time. I also commanded Coltishall whilst that Station Commander was in hospital. One of the worst enemies was the Leeds 'smog' which seemed ever present and the absence of Ground Controlled Approach meant that night flying, particularly, was hazardous and the skills developed in lining up with the runways, and getting home, was astonishing!

As for my personal history, I was commissioned in 1931 — flew Bulldogs — and spent most of my peace-time flying with 605 and 607 Auxiliary Squadrons, teaching them to fly. I went to the Air Ministry in 1937 and left in 1940 to command Manston in September. I had the Northolt Wing in 1941 but was wounded again, coming to Church Fenton via RAF N.I. - leaving there to command 907 Fighter/Bomber Wing in Burma, retiring from the service in 1960.

G. A. L. Manton
October 1987

GROUP CAPTAIN G. C. TOMLINSON OBE DFC

(March 1945 — June 1945)
Now living in Natal — South Africa.

"Reference to my log book tells me that whilst in command at Church Fenton the aircraft on strength were the Mosquito XXX (456 Squadron RAAF— the Spitfire Mk IX and XVI, Oxford, Tiger Moth and Vengeance — operated by 288 Squadron) and a Magister. The Tiger Moth, Oxford and Magister were all communication aircraft, the Spitfires were the day fighters and the Mosquito's had the night-fighter role.

Church Fenton was a sector station in No. 12 Group, Fighter Command, and had a satellite station at Hutton Cranswick where the day fighters were based. It would have had facilities for routine maintenance, but anything other than routine would have been carried out at Church Fenton. My log book states that Group Captain Hawtrey flew me up from Watnall (12 Group headquarters) near Nottingham to Church Fenton on 18 March 1945 in a Dominie aircraft and on the 20 March flew me in the same aircraft to Hutton Cranswick to meet the AOC. of 12 Group — Air Vice Marshal Saul. I well remember Ken Holden (see separate chapter) when he was Station Commander at Hutton Cranswick in 1942 and I note that GAL Manton was here as Station Commander in 1943; we were together in III Squadron in 1933/4 at Hornchurch. I was sorry to hear that John Hawtrey died in 1954, he had a tough time in Iraq during the war when the Iraqi army had the audacity to surround the RAF HQ. and aerodrome at Habbaniyah just west of the Euphrates. There were no fighter aircraft in the country (as I remember) and Hawtry had only Hawker Harts and Hardy's. fitted with bomb racks to retaliate. The Iraqi's were however sufficiently frightened and driven away from Habbaniyah, to relieve the situation.

My tour at Church Fenton was a short, but happy and satisfying one.

G.C. Tomlinson

November 1987

AIR VICE-MARSHALL E. PLUMTREE, C.B., O.B.E., D.F.C.

(November 1945 — February 1946)

"I should like to help with a contribution but I was Station Commander for a very short time — in between three months in hospital and going overseas to do the Staff College course in Haifa, Palestine. At that time the stations main function was to be one of the two centres where repatriated aircrew prisoners-of-war, were sent to be re-habilitated after returning home. There is little I can tell you about that except that my main concern was to help them keep their morale without putting the W.A.A.F. at the near-by Sector Operations Centre, at risk from these sex-starved predators! I am not being frivolous, it was a major task!"

E. Plumtree
August 1987

GROUP CAPTAIN I. S. SMITH, C.B.E., D.F.C.

(January 1952 — July 1953)

"My tenure at Church Fenton was used mainly in the planning of expansion at the Station and the provision of up-to-date facilities. If its a mess, then its all my fault!! One of the more notable events was the extension of the East-West runway. We operated Meteor 8 and were a perfectly normal fighter station in every sense of the description. The majority of the aircrew were national servicemen and within their technical limits were very good indeed. The local people were very kind and generous towards all ranks, as were the Police — of great importance. So our credit memoirs are good"

I. S. Smith
August 1987

GROUP CAPTAIN C. F. GRAY DSO DFC**

(March 1954 — March 1956)

When I arrived for my tour of duty at Church Fenton in March 1954 there were three resident squadrons — all flying Gloster Meteors. There was No. 609 (West Riding) Squadron, No. 72 Squadron and No. 19 Squadron. No. 72 later changed to the Mk 14 NF (night fighter).

We were in 12 Group operating under Northern Sector commanded by Air Commodore V.S. (Tom) Bowling. All the squadrons carried out extensive and successful training programmes, some highlights of which were overseas visits — No. 19 to RAF Jever, No. 72 to RAF Wunstorf and No. 609 for their summer camp to Oldenburg. The Annual Air Defence Exercise "ADEX" was a joint operation in which the tally was 35 destroyed in six actions.

It fell to 609 however, to achieve the greatest honour — the award of the Esher Trophy for the most efficient auxiliary squadron in the Royal Air Force in 1964. Their success was well deserved, as they consistently achieved a high number of flying hours and an excellent standard of efficiency.

A memorable ceremony was held on 31 May, when the Dacre Trophy for 1953 was presented to No. 19 Squadron by the donor, Air Commodore G.B. Dacre CBE DSO. The Air Officer Commander in Chief of Fighter Command, Sir Dermot Boyle, KCVO KBE CB AFC., the Air Officer Commanding No.12 Group, Air Vice Marshal W.J. Crisham and several former members of No. 19 Squadron were present at the ceremony. The Dacre Trophy was awarded annually to the most proficient day fighter squadron in air weapons training, and was in memory of Flying Officer K.F. Dacre, who was killed in action over Germany in 1946.

There were other notable occasions at Church Fenton during the year. On 29th June the station was honoured by the visit of the Supreme Allied Commander in Europe, General Gruenther. On the 11 November, the U.S. Ambassador, Mr. Winthrop Aldridge, arrived at, and departed from Church Fenton when attending a function at Leeds YMCA. He was accompanied by Sir George Martin, the President of the YMCA, and other local dignitaries.

We all enjoyed a successful Battle of Britain Open Day — well attended by the public and adding handsomely in funds to the RAF Benevolent Fund.

Another event I recall was when I allowed the local Hunt to take place on the station, although we were flying that particular day. The 'Master' assured me that there would be no trouble — but forgot to tell the hounds!! They started off in the bomb dump and flew off across the airfield in spite of my instructions, and then decided to come back again, just as the aircraft were coming in — 19 squadron I believe. The C.O. was not amused — nor was I!!

We did not carry a great deal of fuel in those days and I could imagine the headlines in the local paper to read — "Local Hunt strikes aircraft!

C.F. Gray
Waikenae - New Zealand
February 1987

(Photo: Courtesy Group Captain Gray)

***Editors note.** Group Captain Gray ended the war as New Zealand's top scoring fighter pilot with $27^1/_2$ confirmed victories, $13^1/_2$ probables and 12 damaged.*

GROUP CAPTAIN G.D. SISE DSO* DFC*

(March 1958 — July 1959)
now living in New Zealand

My tenure at Church Fenton was highlighted by several notable events. The magnificent Airmens Mess dining hall was completed, and a highly successful Battle of Britain 'At home' Day was held on September 20th, which broke all previous records by attracting twenty five thousand visitors and realising £1000 (a lot of money in those days) in proceeds towards the RAF Benevolent Fund.

Since opening in 1937, the Station had been controlled by Fighter Command, and had been one of the leading fighter stations during this period. This long connection was finally severed in July 1959, when Nos. 19 and 72 Squadrons moved to Leconfield, near Hull, and Church Fenton became a Flying Training Command station. Before the squadrons moved away, much training and activity took place. No. 19 played a leading role in the S.H.A.P.E. (Supreme Headquarters Allied Powers in Europe) defence exercise "Topweight" in April, and a short detachment at Chievers in May, exchanging with No. 8 Squadron, Belgian Air Force. No. 72 Squadron commenced to convert to Javelins in April before leaving for Leconfield. All squadrons and station personnel were heavily involved in the Air Officer Commanding 12 Group's annual inspection, conducted by Air Vice Marshal J.R.A. Embling, CBE DSO.

An event which seemed particularly symbolic took place only a few days before the transfer of the squadrons to Leconfield. On the 11 June the last airworthy Gladiator was flown in by Mr. G. Worral, Chief Test Pilot of the Gloster Aircraft Company. The Gladiator, No. L 8032, appropriately carried No. 72 Squadron markings and was one of the first aircraft to be stationed here. In the afternoon a Meteor, Javelin and the Gladiator were flown over the airfield and Church Fenton's aviational past, and present were merged.

As well as being Station Commander at Church Fenton, I was also responsible for opening up Leconfield and establishing it as a front line fighter station — no mean task. As I recall, the transfer was completed satisfactorily in July 1959. I remained in command at Leconfield until June 1960, when I was posted to H.Q. Far East Air Force at Changi, Singapore.

My memories of Church Fenton are happy ones — a very efficient unit. I also remember the problems we had with poor visibility in winter, due to the notorious 'smog' blown from Leeds.

G. D. Sise
12 February, 1988

Gladiator L. 8032 (Photo: Courtesy Air Portraits)

GROUP CAPTAIN G. J. SOUTH, D.S.O., D.F.C.

(February 1967 — June 1969)

"During my tour of duty the station underwent a period of continuous change, No. 7 F.T.S. having closed the previous year and preparations were made to move the aircrew officers training school from South Cerney. The primary flying school element of this had already arrived and were giving 30 hours basic flying instruction to potential pilots, including some foreign students.

Soon the considerable works service programme needed to accommodate the A.O.T.S. was complete, the move taking place in the early part of 1968. Shortly after, plans were afoot to combine all officers training at a single location, and before I left in June 1969, the A.O.T.C. as such, was wound down preparatory to fusion with O.C.T.U. at Henlow.

Meanwhile, arrangements were being made to take on Naval Initial Flying interests, hitherto carried out at Linton-on-Ouse, one of the main considerations being the siting of the flagpole for the white ensign — a very esential item. The Yorkshire Universities Air Squadron remained a lodger unit, having been expanded around 1968 from the Leeds University Air Squadron and subsuming Hull and Bradford Universities. After I had been at Church Fenton for some months, I made a point of trying to put some historical narrative into the Squadron Operations Record Book (Form 540)".

G. J. South
July 1988

'Hand-over' at Church Fenton June, 1969. Group Captain South (standing) and Group Captain Foster. (Photo: Courtesy Group Captain J. W. Foster)

GROUP CAPTAIN J. W. FOSTER DFC AFC

(June 1969 , July 1971)

My first tour of duty at Church Fenton was in 1952 as Officer Commanding Flying Wing, and later in 1969 as Station Commander. The contrast in roles between the two tours was considerable; from air defence in 1952 on Meteor 8's to primary flying training in 1969 on Chipmunks. Both roles were personally rewarding, and of course the involvement in training both Royal Air Force and Royal Navy pilots was a unique experience for me.

In the latter period of my tenure, the station received the Freedom of the City of Leeds. Another memorable occasion during my appointment was the visit to York by H.M. The Queen and Prince Philip on the 900th Anniversary of the city. The Royal couple landed at Rufforth, our satellite landing ground.

Visit of Her Majesty the Queen Elizabeth and the Duke of Edinburgh
(Group Captain Foster at rear-right)
(Photo: Courtesy Group Captain Foster)

'Handover' of the Station from Group Captain W. G. Abel to Group Captain D. G. F Palmer August 1973.
(Photo: Courtesy Group Captain Palmer)

Arrival of Her Majesty, the Queen Mother — November 1974.
(Photo: Courtesy Group Captain Palmer)

A gathering of No.609 (West Riding) Squadron associates near the Station's Spitfire 'gate guardian', following the Battle of Britain Service —1980
(Photo: Courtesy Air Commodore J. A. Bell)

REACTIVATION OF BASIC FLYING TRAINING (APRIL 1979 — DECEMBER 1980) BY AIR COMMODORE J.A. 'TINKLE' BELL — OBE FBIM

Aggravated by unusually bad weather the Royal Air Force flying training achievement was falling behind the required rate and it was clear that Basic Flying Training School (BFTS) capacity was inadequate to enable the total system to regain lost ground when the weather improved and then keep pace. So a case was formulated to create a third BFTS at Church Fenton to supplement those at Cranwell and Linton-on-Ouse. The airfield was well suited to the operation of Jet Provosts in that role. However, substantial modernisation of Air Traffic Control (ATC) aids and refurbishment of airfield pavements would be needed. Also, the unusually reduced dimensions of certain fundamental facilities (e.g. hangars) at Church Fenton would contain the size of the BFTS to two squadrons, totalling 38 Jet Provosts.

Work on the project progressed at HQ RAF Support Command and in the Ministry of Defence to the point where a submission was made to the then Minister of Defence early in 1978 seeking approval to reactivate Church Fenton at the beginning of April 1979. This would allow the one year deemed to be needed for essential "Get You In" Works Services. Months of further discussion followed and it was not until October 1978 that the RAF obtained Ministerial authority to proceed with the plan.

Early in November I was serving in the Operational Requirements staff of the Ministry of Defence, but was in fact undergoing intensive physiotherapy at RAF Headley Court to speed recovery from a fractured Achilles Tendon, and was surprised to learn that I was to be posted to reactivate, and then command Church Fenton. Terminating treatment forthwith I emerged to hand-over my MOD desk and, make a start on the new task. I was left in no doubt by the AOC (Air Vice Marshal PeterBairsto) that although the date had slipped by seven months the opening date was still to be met.

Therefore the 12 month works Services programme must be intensified and re-ordered to meet the new timescale.

On my initial visit to Church Fenton I met Wing Commander Peter Perry who was to become OC Engineering Wing in the new year on completion of the Staff College course. Straightaway we began to review the concept of the reactivation plan. In general terms we agreed that the flying training task was the primary concern and the pacing item. Thus work on engineering, supporting and administrative facilities would be ordered only in accordance with their relevance to building the operation capability at the required rate.

In January 1979 I left MOD and Wing Commander Perry left Staff College so we both became fully established in our new roles. I proceeded to do the flying refresher course on the Jet Provost which, fortunately, was based at Leeming and operated mainly from Dishforth. When weather prevented flying or I had finished for the day I was able to get to Church Fenton travelling via Linton-on-Ouse to progress the work and obtain knowledge of the flying training business (an unfamiliar field to me). On one such day snow began to fall heavily during the morning and brought flying to a halt at Dishforth. I drove to Linton where already the roads on the camp and through surrounding villages were becoming blocked by snow, and after some work there drove to Fenton. Although the snow was now very thick the camp roads and the public road outside were well swept. This was because the MTO, Warrant Office Squire, with only 2 vehicles at his disposal, had produced a makeshift plough on a Land Rover and cleared the camp roads so efficiently that he was able to send his vehicle out onto the public road as well. It was this enterprising, eager and industrious spirit that was typical of those formative periods. Many airmen (some very junior) found themselves in a position to exercise imagination and conviction and were excited by the rare opportunity to create a new station. They responded well to the occasion and the new station began to take shape.

It was a difficult task for the RAF to provide the 350 or so personnel required to meet the task on opening day rising to 600 or so by the time the second squadron was to be operational in August. I saw people, willing hands and minds, as the pressing need in those early days and told the RAF PMC staff that I would happily accept the greatest possible dilution level of rank and experience to meet the establishment. Thus it was that an unusually young and junior group that built during the early months of 1979.

The youth and inexperience of the station compliment was accompanied by vigour and enthusiasm. All present were stimulated by the sight of the station re-emerging daily. Great credit is due to the Property Services Agency (PSA) team led by Jack Toft, who worked in splendid harmony with me and my staff to ensure that the Part 1 Works Service funds were allocated flexibly so that the workmen were always employed productively regardless of the weather and in accordance with the operational priorities. So, by late March 1979 the essential elements of Ground School including the J P Instrument Trainer were completed and No.1 Squadrons accommodation was ready for use. Engineering Wing had enough bays and facilities to support First line and some Second line servicing and Air Traffic Control though still heavily dependent on Linton for radar, was ready.

Domestically the Sergeants were installed in their Mess but the Officers Mess had been accorded lowest priority and an interim facility in the Airmen's Dining Room building had been created. Married Quarter reactivation and occupation was progressing. The Station Commander's residence, Ulla House, had suffered particularly badly from vandalism and delapidation. Thieves had broken-in, destroyed internal walls and stolen the lead piping which they had concealed. Carcases of dead birds littered the dining room and the garden had been destroyed chemically to prevent overgrowing and hence was barren. However, sufficient repairs were done to permit occupation of the house on 29 March 1979 and the remaining rebuild continued through the following months.

The first eight aircraft were collected from Linton during the last week of March and autonomous operations started in preparation for the formal opening on 2 April. Feeling that the starkness of the playing field area should be relieved by tree development, I accordingly decreed that every major event in the stations life should be marked by planting a horse-chestnut tree, beginning the process on 1 April to denote the stations re-opening, and thereafter each course graduation or VIP visit saw the line of trees extending around the perimeter of the sports pitches area. The AOC arrived on 2 April to be saluted not only by the station personnel on parade but also a formation flypast involving every one of the eight aircraft then on strength.

It had been decided to reactivate the title of No. 7 FTS to give to the new Unit and operations began at once. Indeed they had to because the accummulated backlog in the pilot training system, which had been the root-cause of the reactivation, had reached such proportions that 7 FTS had to play a part in recovering it. Thus on Day One the Unit was told that it was immediately over 200 hours behind its new target line. Buckling-down purposefully all personnel began working hard and constructively as soon as operations began. Courses of student pilots arrived on time and in increasing strength and No. 2 Squadron was commissioned in August 1979.

Instrument Landing System and Precision Approach Radar equipment were installed and calibrated and a surveillance radar was authorised to complete the autonomous suite. The annual SSAFA Air Show became an in-house function again. A year after operations began the refurbish of runways and other operating surfaces began. This lengthy and thorough programme deprived the FTS of one runway at a time but operations continued throughout and the need to provide the contractor with an equipment plus office site and suitable access for his heavy vehicles led to the station acquiring a valuable new crash gate and road to the North of the playing fields.

Refurbishment of the Officers Mess began in 1979 and the new building was progressively occupied culminating in the completion of modernised kitchens in time to have the entire building in use for Christmas of that year. I decided that the original Billiards Room should become the 609 Sqadron Room to commemorate that famous part of Church Fentons history. Suitable memorabilia were used to furbish the room which had a primary role as a television lounge — led by retired Group Captain Stephen Beaumont, then a Sherriff of the West Riding, the 609 Squadron Association were duly welcomed into their new room with a formal reception at which the Earl of Scarborough unveiled a commemorative plaque to continue the involvement of holders of his title with 609 Squadron. Other distinguished visitors during this early period included His Majesty King Hussein of Jordan, who used the airfield during a visit to the ROF in Leeds which was then producing tanks for his army. Security dictated that news and planning for the event be kept on a very tight rein. The same security needs led to such an array of police cars and outriders arriving on the day that the original efforts towards secrecy by the station staff were somewhat nullified.

So the period of my command came to an end in January 1981. The station was then fully operational, all courses had graduated on time and the 200 hour deficit inherited by 7 FTS had become a 300 hour surplus. Whilst all this was going on sport and communal activities had developed and flourished and although at about 700 the station strength was comparatively small, those that were there produced teams and performances, especially in rugby, football, cricket and swimming which put many other larger stations in the shade.

J. A. Bell
August 1988

Editors note: *Since leaving Church Fenton Group Captain Bell has assumed Command of the Officers and Aircrew Selection Centre at RAF Biggin Hill, with the rank of Air Commodore.*

Group Captain Bell in conversation with H. M. King Hussein of Jordan
(Photo: Courtesy of Group Captain J. A. Bell)

GROUP CAPTAIN K. R. BRIGGS

(January 1981 — June 1983)

"My tour at Church Fenton proved to be an eventful one. Following heavy snowfalls in January 1982, the water levels in the catchment area of the River Ouse and the lower vale of York rose to a point to cause severe flooding in towns and cities along the course of the river, including York. Early in the New Year high tides in the Humber estuary and a stiff onshore easterly wind severely checked the run-off of meltwater, resulting in local raising of the water table, great strain on the river banks, and, eventually, wide scale flooding, and to be known as Operation 'Canute'. The severe flooding in Selby was tackled by the Army, and Selby Council requested assistance from RAF Church Fenton, the chief tasks being the filling of sandbags and the distribution of essential supplies in Cawood, Wistow and Selby, where it was eventually necessary to evacuate families. The situation at Selby rapidly became of very serious proportions and 200 soldiers were brought in to supplement the RAF effort. The Army took over control of the joint service operation, and the three-shift system expanded to four. A Flood Relief Cell was established in Station Headquarters and a RAFLO was appointed.

On base, casualties were the families occupying married quarters in Ulla Green, the Station Commander's and the SMO's — who all had to be evacuated. Wide spread flooding of public roads developed, the road from Ulleskelf leading to the Tadcaster road becoming impassable, and transportation in certain areas was impossible, even with Land Rovers. Throughout 6 January sterling work by parties of Student pilots under OC GD Flt. assisted in the sandbagging of all married quarters, the lifting of carpets and the moving of furniture upstairs. Under the Officers Mess itself, the heating plant and the main electricity distribution panel were in danger of flooding but the pumps were able to cope with the situation. DOE maintained a 24 hour presence to monitor conditions in the cellar, and at this stage there was 2 - 3 feet of water in the Station Commander's house, and the station completely isolated by road. On the morning of the 7 January the water level around the station had fallen by 3" or 4" and the Water Authority was satisfied that the crisis had passed. Conditions generally commenced to improve. The PRO kept CPRO up-to-date with flood relief operations, the national Press carried reports of the emergency, and a mention in 'The Times' of 7 January praised the work of the RAF. Local newspapers, TV and radio also featured stories of the RAF in action. In terms of goodwill and morale the gains were enormous, and on balance a most worthwhile exercise.

Astronauts Visit

On 12 May 1982 the Station was honoured by the visit of Colonel James Irwin, the American astronaut. It had been a decade since man had first journeyed to the

Group Captain Briggs briefing the Rt. Hon. Michael Alison M.P. and Major General I.H. Baker G.O.C. North East District on flood operations.
(Photo: Courtesy Group Captain Briggs)

Group Captain Briggs presents the Station crest to Colonel James Irwin
(Photo: Courtesy Group Captain Briggs)

Antony Valentine is given a 'flying lesson' by Group Captain Briggs during the making of T.V. series 'Airline'
(Photo: Courtesy of Yorkshire Evening Press)

Moon in July 1969, and there had been six successful Apollo Lunar Missions to date that had excited the world. Some had asked — "What is the purpose of these missions, what good will they do?" My answer was that man was not impelled to the Moon like the first man to climb Everest — because it was there — nor was it to prove that some political system works better than some other, although there must be some political, economic and military considerations. The answer must be, surely, that by his very nature man has to know.

We were indeed very privileged to welcome Colonel James Irwin among us, when he gave an informal talk to Squadron personnel and an invited audience from the general public.

Colonel Irwin was a Graduate of East High School, Salt Lake City, and received a Bachelor of Science degree from the United States Naval Academy in 1951, followed by a commission in the U.S. Air Force. He graduated from the Experimental Test Pilot School and in 1963 the Aerospace Research Pilots School. In a varied and demanding military career, Colonel Irwin accumulated in excess of 7000 flying hours — over 5300 on jet aircraft.

Following selection as an astronaut by NASA in 1966, he trained as a member of the astronaut support crew for Apollo 10 and a back-up module pilot for Apollo 12. On the Apollo 15 mission, flown from 26 July to 7 August 1971, he served as a lunar module pilot. It was on this mission, the 4th to the Moon, that the lunar buggy was used for the first time. Whilst Colonel Al Worden orbited the Moon in the Command module, Colonel Irwin and the Apollo 15 Commander, Colonel David Scott, explored Hadley Rille and the Appenine Mountains of the Moon. They collected 180 lbs.. of lunar rock, which has aided in scientific research and dating. On this lunar flight, Colonel Irwin logged 295 hours and eleven minutes in space, of which 19 hours and 46 minutes were spent outside the module. Colonel Irwin described his experience on the Moon and added "All astronauts say they are not frightened — but we are human — and I certainly felt my heart beating faster than it should do at times". Colonel Irwin has since retired and founded a religous organisation.

"Airline"

Within a few days of my tour of duty at Church Fenton as Station Commander I was face-to-face with a rival 'CO' in the shape of actor James Taylor, taking part in the Independent Television production "Airline" — a 13 part serial which traced the progress of an RAF NCO pilot who starts his own airline. Vintage Dakota aircraft and a Tiger Moth took part, the star of the series being ex-Sandbagger Roy Marsden, supported by actor Antony Valentine. Across the board attention to detail was the order to create authenticity for aircraft, uniforms and scenes.

K. R. Briggs
June 1988

GROUP CAPTAIN R. HUMPHREY ADC

(June 1983 — July 1985)

The first picture that comes to mind about RAF Church Fenton is the air of the carnival! Perhaps this is a surprising recollection about a small, quiet Royal Air Force station set in a peaceful part of Yorkshire. However this is easily explainable, as I arrived to take over the Station as the marquees and side-shows were being built for the 1983 SSAFA Air Day, and I moved on as they were being dismantled after the 1985 Display. In between, the 2 years in command passed remarkably quickly, and the memories are already kaleidoscopic.

There were no truly remarkable events during my stay, an indication that the Station completed its flying training task efficiently and without fanfares. There was a great deal of activity on the work service side, continuing to repair the ravages of the period that the station was closed.

Particularly pleasing was the large scale renovation of Airmen's Married Quarters and the continuous programme of maintaining the general appearance of the station and its reputation for being one of the most attractive RAF bases in the country. Whilst the basic role of the station did not change during my time in command its task increased significantly with the redeployment of the Refresher Flying Squadron from RAF Leeming in 1984, at last RAF Church Fenton was back to full capacity.

The abiding memory of my tour was the way that we at Church Fenton were made to feel an integral part of the local scene. We all gained much from this. Newcomers to the station felt that they were welcome members of the community, and I know that a great deal of 'behind the scenes' assistance by station people and their wives went into helping the local area, with charity functions etc. But involvement in the local community is not just doing "good works" — it is also a matter of joining in. Perhaps a good example of this was the long-running Mastermind quiz competition with teams from most sections and the local pubs battling it out for months on end. I shall never forget the humiliated looks on the faces of one of the all-male pub teams when the Wives Club made a clean sweep of the 'sports' round. Equally, I shall never forget either, the same looks on the Wives Club team when the opposition immediately responded with maximum points for "Art and Classical Music". This was not the Yorkshire that I read about as a Southerner!

What highlights? Visits by HRH Prince Michael of Kent to the SSAFA Air Day in 1984; by the Duchess of Gloucester in 1985 stand out. Also the many formal and informal events resulting from our civic ties with Leeds, and following the change of shape of local Government in Yorkshire, the developing ties with Selby District. Meeting on and off duty the unnumerable groups and organisations that shared common interests; the ATC Squadron; the RAFA and Aircrew Association branches; CCF units — spread far and wide throughout Yorkshire. All of these gave me a unique insight into the strength of support for the Royal Air Force which I always assumed, and hoped, existed but had never experienced at first hand. Above all, the innumerable visits by young and old alike to see the station in action, left me with the last impression that Church Fenton 'belongs' to Yorkshire past, present and future.

R. Humphrey
July 1988

Visit of H. R. H. Michael of Kent, with Group Captain R. Humphrey.
SSAFA AIR DISPLAY - 1984
(Photo: Courtesy of Administration Wing, RAF Church Fenton)
(Reproduced by permission of the Controller - Her Majesty's Stationery Office)

GROUP CAPTAIN C. J. STURT OBE MBIM

Tour No 1 — (May 1963 to March 1966)
Tour No 2 — (July 1985 to August 1987)

I first served at Church Fenton from May 1963 to March 1966, having recently left Central Flying School at Little Rissington as a B2 — Qualified Flying Instructor. No. 7 Flying Training School had just been formed, equipped with the Jet Provost Mks. 3 and 4, and I arrived about the time that No. 2 course graduated, and left with the arrival of No.17 course — the last course of that era.

My time at Church Fenton consequently spanned very nearly the whole of No. 7 FTS operation in the early sixties. It was, of course, a basic training school, after which students progressed to the Gnat (fast jet) or Varsity (multi-engined) for their advanced training. However, in those days pilots graduated with their 'wings' from the FTS, unlike a few years previously and the system today.

I was attached to No.1 Squadron which had three squadron commanders at this time — Gerry Eades, Bill Waite and Mike Armitage, now Air Chief Marshal and Commandant at Royal College of Defence Studies (RCDS). This is the most senior course, having a duration of twelve months, to which the Services send students of Group Captain/Air Commodore level. Police, Civil Servants and foreigners also attend.

Our students were mainly RAF but we did have Kuwaitis and Syrians also. The station was small and compact, with good morale and a lively mess life. (It was in the officers mess that I was first introduced to a Miss Ann Parsons and whom I married in 1964 — our first son, Simon, was born in York Hospital in 1965)

On my return to Church Fenton as Station Commander some nineteen years later it seemed as though time had almost stood still. The cherry trees lining the entrance to the officers mess were still there, and apart from a new building — now a community centre — there seemed little change to the infrastructure. The station still had the same friendly atmosphere of twenty years previously and the raw material — keen young students — were, as ever, ready to take up the challenge of becoming RAF pilots.

My impressions, having once again departed from Church Fenton, is that the product leaving the FTS is as good a pilot as before but very much better at flying low level — the training is better as the emphasis has focused primarily in the RAF to this area. No longer do they get their 'wings' on graduation — the wheel having turned a full circle — and no longer is there the difficulty of not graduating pilots who should plainly have failed (In the early fifties we went through a patch — disastrous in my view — when the then Commander-in-Chief, Flying Training Command, decreed that there were no bad students — only bad QFI's)

Mess life has, however, changed and no longer seems to be the focal point of the

Officers as I remember before. This, I suspect, reflects the changing aspirations of our youth over the years, and the fact that, relatively speaking, they are better paid now than before.

As previously, the station has integrated well into the local community, and Ann and I certainly made many friends in the area during our recent tour.

I could go on much longer, but in the interests of reasonable brevity I will cut short here

Charles Sturt
October 1988

The Prime Minister the Rt. Hon Margaret Thatcher M.P. with Group Captain C. J. Sturt OBE.MBIM. arriving at Church Fenton en route for her visit to Leeds. February 1987 (Reproduced by permission of the Controller - Her Majesty's Stationery Office)

THE STATION BADGE

The Station Badge was presented in June 1953 and depicts three tilting Spears in front of a Yorkshire Rose. Through an allusion to the former fighter role of the Station, the badge and motto remain appropriate to the need for the swift and efficient training of aircrew.

(Reproduced by permission of the Controller - Her Majesty's Stationery Office)

RESIDENT SQUADRONS AT CHURCH FENTON

(Reproduced by permission of the Controller - Her Majesty's Stationery Office)

Squadron No.	Designation	Arrive - Depart	Aircraft Used
72	RN	1.6.37 - 16.10.39	Gladiator Spitfire
		14.1.40 - 2.3.40	Spitfire
		9.5.53 - 29.6.59	Meteor Javelin
213	AK	1.7.37 - 18.5.38	Gauntlet
64	SH	18.5.38 - 11.5.40	Demon Blenheim Spitfire
242	LE	30.10.39 - 21.5.40	Magister Tiger Moth
			Blenheim Battle
			Hurricane Master
234	AZ	22.5.40 - 18.6.40	Spitfire
		18.7.43 - 5.8.43	Spitfire
		28.12.43 - 29.1.44	Spitfire
249	GN	16.5.40 - 18.5.40	Hurricane
		9.7.40 - 14.7.40	Hurricane
87	LK	26.5.40 - 5.7.40	Hurricane
73	TP	22.6.40 - 21.9.40	Spitfire
306	UZ	28.8.40 - 4.11.40	Hurricane
85	VY	5.9.40 - 23.10.40	Hurricane
		18.9.57 - 30.11.58	Meteor
71	XR	19.9.40 - 23.11.40	Buffalo Hurricane
54 (OTU)		25.11.40 - 6.5.42	Blenheim Defiant
			Oxford Beaufighter
25	ZK	16.5.42 - 19.12.43	Beaufighter Mosquito
488	ME	25.6.42 - 1.9.42	Beaufighter
600	BQ	1.9.42 - 14.11.42	Beaufighter
183	HF	1.11.42 - 25.3.43	Typhoon
308	ZF	29.4.43 - 4.7.43	Spitfire
26	XC	20.7.43 - 27.12.43	Mustang
96	ZJ	6.8.43 - 6.9.43	Beaufighter Mosquito
125	VA	24.4.45 - 20.11.43	Mosquito
264	PS	19.12.43 - 5.5.44	Mosquito
		20.11.45 - 18.7.46	Mosquito

Squadron No.	Designation	Arrive - Depart	Aircraft Used
288		20.11.44 - 27.6.45	Spitfire Hurricane Oxford Tiger Moth Vengeance
(A detachment was at Church Fenton from December 1943)			
124	ON	18.3.44 - 22.4.44	Spitfire
604	NG	25.4.44 - 2.5.44	Mosquito Beaufighter
307	EW	4.5.44 - 29.1.45	Mosquito
456	RX	31.12.44 - 16.3.45	Mosquito Beaufighter Blenheim Defiant Oxford
68	WM	15.3.45 - 20.4.45	Mosquito
129	DV	28.6.46 - 31.8.46	Spitfire
	(renumbered 257 Sqn. 1.9.46)		
609	PR	10.5.46 - 5.11.46 6.10.50 - 10.3.57	Mosquito Oxford Spitfire Harvard Vampire Meteor
263	HE	2.4.46 - 2.5.47	Meteor
257	EE	1.9.46 - 15.4.47	Spitfire Meteor
19	QV	24.4.47 - 26.6.59	Hornet Oxford Mosquito Meteor Hunter Lightning
41	EB	18.4.47 - 29.3.51	Spitfire Hornet Meteor Mosquito Javelin Phantom Jaguar
Leeds University Air Squadron		29.10.59 - 6.3.62 29.4.66 - 15.3.69	Chipmunk Bulldog
23 Group Command Flight		14.9.59 - 12.2.62	Anson Meteor Provost
7 Basic Flying Training School (BFTS)		13.3.62 - 30.11.66 (disbanded)	Provost Jet Provost Vampire
(reactivated)		2.4.79 - date	Jet Provost

2b

STATION TASK AND ORGANISATION

R.A.F. Church Fenton is a station in Support Command, housing No. 7 Flying Training School. The Station Task and organisation is as follows:

FLYING WING

Task: Flying Wing is commanded by the Chief Instructor who is a Wing Commander, and consists of two basic flying training squadrons, Headquarters Squadron, Ground Instructional Squadron, Air Traffic Control Squadron, a General Service Training Office and a Wing Headquarters. Flying Wing has two tasks, the greater of which is to train young officers as pilots to prepare them for advanced training on either the Hawk, the Jetstream or the Gazelle. The second task is to provide refresher flying training to pilots returning to flying duties after one or more Ground Tours.

ENGINEERING WING

Task: Engineering Wing works in close liaison with Flying Wing, to ensure the provision of sufficient serviceable aircraft throughout the flying day to meet the station task.
The Wing consists of a Headquarters Flight, Aircraft Servicing Squadron, Support Squadron and the Electrical Engineering Squadron.

ADMINISTRATIVE WING

Task: Administrative Wing is responsible for the administration, management and support of the station as a whole, and comprises Station Services Squadron, Personnel Management Squadron and Regt. Flt.

3

No. 72 SQUADRON

Squadron Crest: a Swift in flight. *Motto: "Swift"*

No. 72 Squadron was formed at Netheravon in July, 1917, and trained scout pilots for service in the Middle East in November of that year. The Squadron was based at Basra in March, 1918, equipped with five different types of aircraft, which must have made the supply and servicing problem extremely difficult.

Shortly after its arrival the Squadron was split up into three flights, each of which was attached to an Army Corps where they carried out the vital, if unglamourous work of Army co-operation. In this connection "B" flight was called upon to assist General Dunsterville's expedition, "Dunster Force".

In October, 1918, the operation which was to result in the surrender of the Turkish Forces on the Tigris Front was launched and the Squadron harried the enemy with bombs and machine guns over a large area. Mosul was occupied in November, 1918, and the Squadron then returned to Bagdad where it was disbanded in December,1919.

During its brief existence the Squadron pilots flew D. H. 4's, S. E. 5's, Spads, Martynside Scouts and Bristol Monoplanes over half a million square miles of desert country in extreme conditions, where the natives were definitely unfriendly.

No. 72 Squadron reformed at Tangmere in February, 1937, and were equipped with Gloster Gladiators, which were very popular with the pilots. Later that year the association with Church Fenton began; an association which has continued to the present day, with the exception of the war years. In April, 1939, the first Spitfires arrived, greatly increasing the striking power of the Squadron, and these aircraft saw action shortly after the start of the war, when in October, 1939, a patrol shot down two Heinkel bombers without loss. From this time until August, 1942, the Squadron was engaged on normal fighter operations, including the Battle of Britain in which it fought with great distinction.

In November, 1942, the unit moved to the Middle East where it was engaged in operations in the Western Desert, flying offensive sweeps and carrying out escort duties. During this period the Squadron scored 53 enemy aircraft destroyed, 13 probables and 50 damaged; more than any other squadron in the campaign. In June,

1943, after a short rest, a move was made to Malta to fly sweeps over Sicily, where in one day 8 enemy aircraft were destroyed without loss.

During May, 1945, the Squadron moved to Klagenfurt, in Austria, and carried out tactical reconnaissance for the Jugoslav Partisans.

In all operations "72" destroyed 248 enemy aircraft, claimed 67 as probables and damaged 181; this being the second highest score in Fighter Command.

They were disbanded in Italy in 1945 but were reformed two years later at Odiham, being equipped with Vampire jet aircraft, as a day fighter squadron. The Vampires were eventually replaced by Meteors, and after a move to North Weald the Squadron came back to its old home at Church Fenton in 1953.

AN ERK'S EYE VIEW OF CHURCH FENTON IN 1939

"In the summer of 1939, there were two squadrons at Church Fenton; one was 72 Squadron which was being re-equipped with Spitfires, but still had one flight of Gladiators. The other was 64 Squadron which had short-nosed Blenheim fighters, armed with a cluster of four Brownings under the nose, and one Vickers 'K' in the dorsal turret.

During this period I was a Flight Mechanic (Airframes) - with 72 Squadron, which was commanded by Squadron Leader Lees.

Being only eighteen and an AC2, I only had contact with one officer, namely Flying Officer Sheen, who was an Australian and still wore the navy blue uniform of the Australian Air Force. I was placed in charge of his aircraft's airframe together with Flight Mechanic (Engines) AC2 Rockliff from Barnsley, who was responsible for the engine. The aircraft number was K9959 and I seem to remember the identification letters were RN-R. The 'Chiefy' in charge of maintenance in 'B' Flight, was Flight Sergeant Laws.

There were no runways in those days and when the ground was heavy and soggy with rain, the wheels of the Spitfires would sink into the soft ground when taxiing and prevent them rolling forward. This coupled with the pull of that big heavy engine at the front, would tip the old Spitfire onto it's nose, bending the aluminium propeller blades forward like the petals of a flower.

When the bent propellers were taken off, they were placed in one corner of the hangar which we called the 'Octopus Den', because that is exactly what it looked like with all the bent blades coiled upwards like tentacles.

The surnames of the Pilots of 72 Squadron at that time were Pilot Officer Elsdon, Pilot Officer Elsworth, Pilot Officer Pigg, Pilot Officer James Nicolson (see later chapter), Flight Officer Graham, Sergeant Norfolk, Sergeant Dowthwaite — there were many more but those are the ones that I remember.

At that time the 72 Squadron had the job of testing a new three bladed wooden propeller made by JABLO. It was fitted to one of the Gladiators and was the most wonderful piece of workmanship I had ever seen. Made of stained mahogany and polished to a glass finish, it seemed a shame to use it. But I understand it was of great practical use and the Gladiator could almost hang on it at full bore.

The ground crews used to make a bee-line to the nearest towns on evenings and week-ends. When returning late at night they would wind their way back to camp in a long line along the country lanes, many of them calling at the little wooden hut shop for a Lyon's Individual Fruit Pie which they munched on the second half of the journey to the guard room in time for the golden hour 23-59. . . . .

The 3 September changed all that, all personnel were confined to camp, the aircraft were dispersed around the perimeter and guards placed on them (the Home Guard usually) and it suddenly became quite dangerous going to work in the early hours when it was dark, because they always had one up the spout. It was wise when challenged by a "Who goes there" to answer "Friend" loudly and quickly. Sometimes in the blackness it was not uncommon to hear three Erks answer the one challenge.

During October a Bombay troop carrier landed and without any warning, the ground crew of 72 Squadron were whisked off to Drem in East Lothian. The Spitfires arrived shortly after and we set about establishing ourselves there. This happened many times to 72 Squadron, as we were a mobile unit and had to be able to move within a few hours."

72 Squadron went on to distinguish itself at Biggin Hill during the Battle of Britain. AC Price later re-mustered to Air Crew through the Empire Air Training scheme in Canada and completed a tour of thirty-two operations as a Pilot on Lancasters, and was later awarded the DFC for conspicuous skill and gallantry in landing a disabled Lancaster bomber armed with a full complement of bombs.

George Albert Price DFC
August 1988

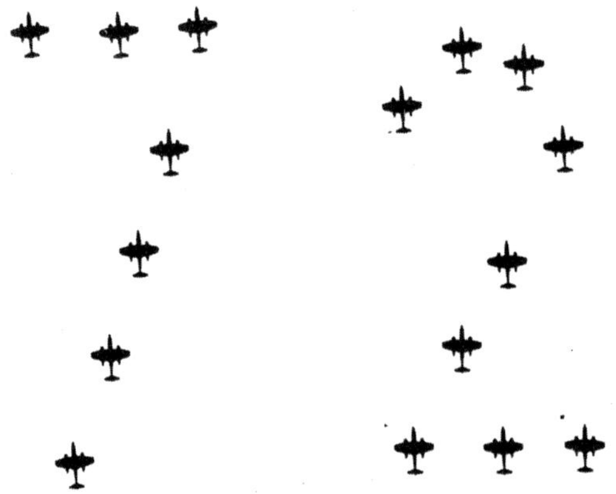

Meteor XIV's of No. 72 Squadron on formation practice.
(Photo: Courtesy Group Captain C. F. Gray DSO. DFC.++)

4

THE STORY OF 249 SQUADRON

(Motto: PUGNIS ET CACIBUS - (With Fists and Heels)

No. 249 Squadron was formed in the last few months of the First World War and was engaged in anti-submarine work off the Scottish coast, operating from its base in Dundee, from August 1918 until it was disbanded in 1919. After fourteen months of existence, the Squadron had to wait until the next major crisis in Britain's modern history, before the story was continued.

On the 16 May, 1940, No. 249 was re-formed at Church Fenton. The pattern for re-armament within the Royal Air Force was that new squadrons were hurriedly being formed to meet the challenge from the Luftwaffe. No sooner were they trained and considered operational than they were pitchforked into the line to face the 'fury and might' of the enemy.

Typical of this process was No. 249 Squadron, re-formed, commanded and made operational within six weeks by Squadron Leader John Grandy (Marshal of the Royal Air Force, Sir John Grandy).

Its experience is specially significant because from this momentarily-created base, it was to develop into one of the most successful wartime fighter squadrons in the Royal Air Force.

Also unique among its distinctions, was the award of the Victoria Cross, the supreme wartime honour, to one of its members, Flight Lieutenant J. B. Nicolson, on 16 August 1940. (More of this in a separate chapter). Grandy has recorded his recollections of the squadron in those tumultuous summer and autumn days of 1940.

"On 16 May 1940, I received orders to re-form No. 249 Day Fighter Squadron at Church Fenton, Yorkshire. Armed with Spitfires we rapidly got going on these brand-new, exciting aircraft. We were not to enjoy them for long — the memory dims. Spitfire production may have lagged for some reason or, more likely, our aircraft were needed to strengthen already operational squadrons in the south. Whatever the cause, away they all went and we were sad — but not for long. There was no time for that. We were re-equipped with Hurricanes, also brand-new and fitted with the Rotol three-bladed variable pitch propeller, and there soon developed a great enthusiasm for these excellent sturdy fighters.

My orders from AOC 13 Group were that, despite the change of aircraft, I had to get 249 operational within one month.

We flew hard and we flew thoroughly, I was ably supported in our training task by my two flight commanders, Flight Lieutenants Kellett and Barton and a superb ground staff to keep us in the air; Flight Sergeant Bennett was a tower of strength in this work.

At the end of June, Air Commodore Nicholas, SASO, 13 Group inspected the squadron and pronounced us operational. There was an important ingredient in this success. I knew that our squadron training task would be made infinitely the easier if I had good pilots. I wanted the best I could find. Fortunately, I had been Adjutant and Flying Instructor of the University of London Air Squadron in 1936/37. I knew all the University air squadrons and had a high regard for their members' abilities. Many of the undergraduates whom we had been training to 'ab initio' standards on the Avro Tutor in 1936/37 had by 1940 joined up and were doing their advanced training on Hawker Harts (and I believe Furies) at Cranwell and other training schools. So, I managed to "fix it" that I got a very good share of the best that was going in terms of flying ability, guts and skill.

This was to stand us in very good stead. The reputation which the squadron built during the Battle of Britain was to remain with it throughout the war. New pilots, as they arrived both in the UK and, later on, when 249 Squadron had moved to the Mediterranean, very soon became aware that the they had to fly and operate to a pretty high standard if they were to make the grade.

It is not easy to record accurate impressions of events, however momentous, which took place almost fifty years ago. Of one thing one can be sure: all the emotions were there, excitement boredom, eagerness, fear. We knew that an invasion was on the cards. We were, I believe, spurred on by this, and by anger at the Luftwaffe's attempt to fly in our sky, to bomb our country, our people. Victory was not in sight, yet the thought of defeat never entered our young minds.

On such foundations was the wartime history of 249 Squadron to be built"

On 18 May the Squadron left for Leconfield returning to Church Fenton on the 8 July, finally leaving for Boscombe Down on the 14 August to serve with distinction in the Battle of Britain and the Battle of Malta.

After the war, No. 249 took part in the Egyptian (Suez) campaign, flying D. H. Venom F.B.1 and F.B.4. Shortly afterwards the Squadron moved to Sharjah to operate against the Imam rebels in the Trucial States. In October 1957, the unit once again disbanded — and re-formed — this time at Akrotiri, Cyprus, re-equipped with Canberra B2, B6 and B16 aircraft. After a further twelve years in the area, the Squadron was disbanded on 24 February 1969.

Squadron Leader John Grandy with Air-Vice Marshall Saul and Air Marshall Sholto Douglas (Photo: Courtesy 609 Collection)

5

THE EAGLE SQUADRONS

Although this book is primarily concerned with the squadrons and events at Church Fenton, it would be incomplete without mention of the overall picture, background and motivation behind the American involvement in the early days of World War II — 'Hitler's War'.

Mr. Charles Sweeny — a staunch friend of Britain and founder of the Eagle Squadrons — re-captures the mood of the times and recalls:-

"In 1940 and 1941 — 'Our darkest hour" as Winston Churchill termed it — a number of young Americans came to England determined to defend with their lives the principles of freedom and democracy. They were banded together within the Eagle Fighter Squadrons of the Royal Air Force and set up a very distinguished record. When the United States came into the war in December 1941, they were absorbed into the United States Army Air Force as the 4th Fighter Group and again were extremely successful. Amongst the names that spring to mind are Tobin, Mamedoff and Keough, the first three volunteers, followed shortly afterwards by Peterson, Goodson and Blakesly. These boys, as well as others, will live forever wherever brave men and their deeds are recounted".

THE EAGLES WAR

By the Autumn of 1940, with the Royal Air Force still sorely pressed by the German Luftwaffe, the beginning of a stream of American pilots was moving to swell the ranks of the RAF. — while the organisational procedures were being implemented to create the fighting unit.

That Yanks were on the way was due in no small part to the foresight of the British and Canadians, and the tenacity of some concerned Americans. From the first days of war, a mechanism had been in action to screen potential pilots from the many American volunteers that were expected to try for the Royal Air Force and Royal Canadian Air Force. To handle the operation, RCAF Vice Marshal Billy Bishop had approached Clayton Knight, an American who had flown for the Royal Flying Corps in World War I and who was an artist held in particular respect by aviation enthusiastics for his excellent depictions of aerial warfare.

Knight and an influential and hardworking friend, Homer Smith, set up interviewing and screening operations in various cities, at first clandestinely, and then openly. By the time of Pearl Harbour, the Knight Committee had handled 50,000 applications and had cleared 6,700 pilots for service with the RAF and RCAF.

Charles Sweeny goes on to exlain:-

"As far as the Eagle Squadrons are concerned, they really had nothing to do with the Clayton Committee, which was formed by Clayton Knight with the co-operation of the Empire Training Scheme, headed by Air Marshal Billy Bishop VC. Billy Bishop was an old family friend and the idea was that the Clayton Committee would facilitate volunteers to the Allied Air Forces. He worked in conjunction with a cousin of mine, Homer Smith, and by agreement with Harold Balfour, who was then Under Secretary of State for Air.

We agreed that volunteers to the RCAF would either go directly into the RCAF or, if they had 30 hours flying, they could go to England and join the Eagle Squadrons.

The Eagle Squadrons were formed by me in May 1940 after conversations with Churchill, Beaverbrook, and a meeting with the Air Staff. They listened to my suggestions and in due course, I received a letter saying that 71 Squadron had been activated by the Royal Air Force to house the Eagle Squadrons, but in the first instance they wanted 25 pilots and 25 pilots in reserve, before they became operational. The same formula was followed up when the other two squadrons were formed, but of course, they were acting through the Empire Training Scheme which provided a steady flow of pilots.

I managed to get my Uncle Charles as the non-flying executive of the Eagle Squadrons with the rank of Group Captain. The trouble was that he was a boyhood hero of mine, and of course, had no real interest whatever in an air squadron. In fact, when he arrived with Bill Taylor and my brother Bobby, and I informed him of the formation of the Eagle Squadrons, with the rank of Group Captain for him, he absolutely refused the appointment. He thought he was coming over to command a battalion of American volunteers in an army unit, and my brother and I had to persuade him to accept the position which I had negotiated for him. He really didn't last very long and wasn't at all interested, and eventually he retired to civilian life after a few months of effort.

The first three volunteers were Mamedoff, Tobin and Keough. It was rather funny, because the Air Ministry rang me, about the day after I received a letter saying that the Air Staff had formed the Eagle Squadrons and were prepared to support them, to announce to me that my first three volunteers had arrived. I remember I was amazed and I took them to White's Club for dinner where they recounted to me all the vicissitudes which had plagued them in their efforts, first to join the French Forces and secondly, to get to England. It was all very interesting at that particular time and as more volunteers came along it was obviously going to be successful.

I remember flying up with Harold Balfour to Kirton Lindsey, where we reviewed the newly formed Eagle Squadron.

Oddly enough, the name 'Eagle Squadron' was suggested to me by my father and I took the shoulder patch from my American passport and added the initials 'E.S.'.

This is a brief resume of what actually happened in the formation of the Eagle Squadrons and, as I say, it was a very interesting time. Eventually, we had over 240 volunteers and some 80 were lost in action, which as you realise, is a very high percentage"

In anticipation of the flow of recruits from across the Atlantic, the Royal Air Force established the All-American Fighter Squadron at Church Fenton — whose motto was appropriately, 'First Out of the Eyrie' — and given No. 71 Squadron.

Historically, there had been a No. 71 Squadron back in World War I, flying Sopwith Camels and Snipes. This was the fourth British Squadron to be formed from Australian volunteers and it achieved the highest score of victories, with seventy-six German aircraft destroyed. It was disbanded in 1918.

The first men to form the embryo squadron were Eugene Tobin, of Los Angeles, Andrew Mamedoff of Connecticut, and Vernon Keough of Brooklyn, New York. They were escapees from France and on reaching England, the three had enrolled in Squadron No. 609 (West Riding of Yorkshire) then based at Warmwell, in Dorset.

In a month of action there, Tobin destroyed three enemy air craft and damaged two; Mamedoff and Keough shared in shooting down a Dornier bomber. Then they were sent to Church Fenton.

Well-liked and admired, the Americans had come to fight long before their own country was involved in the war. Eugene Tobin (called 'Red' because of his hair) and Andrew Mamedoff (of Russian extraction), had originally enlisted for the war in Finland — despite the fact that Tobin's previous flying experience was limited to some 200 hours in light aircraft.

Mamedoff had been a barnstormer, using his own aircraft. They came to Europe with Vernon Keough to join the French Armée de L'Air, but arrived too late, and were fortunate to get away on the last ship leaving France.

They were a colourful, wise-cracking threesome. 'Red' had a delightfully casual outlook and manner; Andy loved to gamble; 'Shorty' was an ex-professional parachute jumper and at 4'10" was the shortest pilot in the Royal Air Force — he needed cushions to elevate him to look out of his Spitfire windscreen.

The Squadron was honoured by a visit from Air Commodore, The Duke of Kent. 'Shorty's reaction was —"say, what do we call this guy — Dook?". He was informed that "Sir" would suffice and he later had a lengthy chat with the 'Dook'. Full of courage, as they undoubtably were, they were also inexperienced in aerial combat and the Commanding Officer, Squadron Leader H. S. Darley, wisely restricted their operational roles until 'they could look after themselves'. 1

Later they all acquitted themselves with distinction, but were, regrettably, all later killed in action.

SQUADRON FORMATION IN A BORROWED WAR

"From, six at first we soon became
A Grecian clan of wondrous fame".
(College fraternity song)

There were only three at first in No. 71 Eagle Squadron not six. They became a Royal Air Force — not a Grecian clan of wondrous fame.

The Squadron Operation Book records:-
19.9.40 Arrival of the first group of pilots for Eagle Squadron No. 71 — Pilot Officer G. Tobin; Pilot Officer A. Mamedoff and Pilot Officer V. C. Keough having been posted from No. 609 Squadron each having had about 50 hours operational service on Spitfires.

The second entry, ten days later reads:-
29.9.40 Squadron Leader Walter Myers Churchill, DSO DFC arrived from command of No. 605 Squadron at Croydon.

Arrival of Pilot Officer F. H. Tann, Adjutant. Pilot Officer Donahue arrived from No. 64 Squadron, having seen a good deal of service, having been shot down and quite badly burned"

In RAF terminology, Pilot Officer is the equivalent of an American Second Lieutenant; Flying Officer is equivalent to First Lieutenant; Flight Lieutenant equivalent to Captain; Squadron Leader equivalent to Major and Wing Commander equivalent to Colonel.

Arthur Gerald Donahue, the fourth American pilot to join No. 71 Squadron, arrived at Church Fenton on 29 September 1940 with Englishman Walter Churchill, who assumed command of the new unit, and Donahue, fresh off the family farm near St. Charles, Minnesota, had been assigned to the all-British 64 Squadron upon his arrival in England on 4 August 1940. The very next day he tangled with enemy fighters and barely managed to bring his crippled Spitfire home. On 8 August 1940 the Squadron intercepted thirty raiding German planes, and Donahue damaged and possibly destroyed a Messerschmitt.

In his third engagement, four days later, cannon shells tore away his elevator and control cables, and incendiary bullets set his Spitfire aflame. Donahue bailed out, and British soldiers later found him with his legs, hands and face badly burned. In little more than a week of action with the RAF, Donahue had become something of a war hero in England. Seven weeks later, the durable youngster was out of the hospital and in the new Eagle Squadron.

On 4 October, more men joined the new Squadron; two American Pilot Officers, Philip H. "Zeke" Leckrone, a farm youth from Salem, Illinois and Luke Elbert Allen from Ignacio, Colorado. Two British officers, Flight Lieutenant's George A. Brown and R. C. Wilkinson arrived, assigned later to lead two of the Squadron's three flights, followed by more Americans, Squadron Leader William Erwin Gibson Taylor, a former aeronautical engineering student at New York University, and Robert V. "Bobby" Sweeny of London, winner three years previously of the British amateur golfing title.

Bill Taylor had been an ensign in the US Navy and was a First Lieutenant in the US Marines, on vacation in England in September 1939. Four days before the outbreak of war, he received permission to join the Royal Navy for which he flew from the British aircraft carriers "Glorious" and "Furious", in operations off Norway.

Taylor was in Washington in the Spring of 1940, temporarily attached to the British Naval Attache and to the British Purchasing Commission in New York City, when he made the acqaintance of Bobby Sweeny and Colonel Charles Sweeny. The Sweeny's decided that the tall, aristocratic looking and highly qualified flier, would be ideal to command an All-American RAF unit. Bobby's brother, Charles, who was in London, concurred.

The three men went to England by ship together. Taylor recalled that in front of newsreel cameras at a large press occasion at Northolt, he was transferred to the RAF and made a Squadron Leader. Bobby Sweeny was commissioned a Pilot Officer and given the title of Adjutant of 71 Squadron, whilst Charles Sweeny was made a Group Captain, and appointed Honorary Commander.

According to Taylor, *"The three of us went to pay our respects to the Air Officer Commanding 12 Group, Air Vice-Marshal Trafford Leigh-Mallory — who received us coolly!*

He said that he was very strongly opposed to having a Squadron of Americans, having had experience with them in the First World War and found that individually they were charming, but as a group they were completely undisciplined. I kept my mouth shut except to say that the Squadron's formation was already a fait accompli"

Taylor and Bobby Sweeny joined 71 Squadron at Church Fenton on 4 October 1940, only to discover that Squadron Leader Churchill and Flying Officer Tann, both British, had arrived five days earlier and had taken office as Squadron Commander and Squadron Adjutant, respectively. Taylor said that he was not disturbed by the apparent duplication in command, since it was natural to have RAF officers indoctrinate Americans.

These would-be combat pilots had no equipment, not even one aeroplane to fly. The single aircraft that had been provided for them, a Miles Magister trainer, was unserviceable.

Donahue could stand the enforced inaction no longer, and on 23 October 1940, at his own request, he was allowed to go back to an all-British Squadron that had flyable equipment. The 71 Squadron Operations Record Book entry on that date:-

"At this point the entire personnel were completely browned-off because of the enormous lack of interest in the equipping of the squadron with any aircraft".

Just twenty-four hours later, the Operations Record Book read:

"This was a momentous day, for three Brewster American fighters arrived and the pall lifted".

Another Brewster arrived the following day. Enthusiasm over the single-seat, single-engine Brewster F_2A Buffalo however faded quickly. Developed for the US Navy and not well adapted to land-based operations, the plane had a disconcerting tendency to float on its approach to landing. The pilots also found the Brewster to be slow and limited in attitude capability, too lightly armed and too vulnerable structurally for use in combat. The plane also lacked self-sealing fuel tanks. The tanks were an intergral part of the wing, and a single bullet in the wing could set it afire. There were no spare parts available, and the plane's sheet metal and hidden parts made maintenance difficult.

Concluding that the small aircraft were deathtraps, Churchill decided quietly to get rid of them, the three that were still operational. He ordered Wilkinson, Mamedoff and Tobin to take off in the Brewsters and to land without locking the tail wheels.

This startled the pilots, because standard instructions for the Buffalo were to make sure that the tail wheel was locked for the landing, but they soon had a fair idea of what Churchill had in mind. A Buffalo landing with its tail wheel unlocked, usually would ground loop. In a ground loop at high speed, the plane would tilt to one side and dig its wingtip into the ground, and a damaged wing meant a fuel tank wrecked virtually beyond repair.

Each pilot landed with the tail wheel unlocked, as ordered. The automatic result was a series of devastating ground loops. In one stroke the entire Brewster "squadron" was wiped out, but no one was hurt. The RAF remained discreetly silent about the disappearance of the American Buffalo from its inventory.

Before deciding to eliminate the Brewsters, Walter Churchill had received assurance that better equipment, the Hawker Hurricane I, was on its way. The first British low-wing fighter and the first operational RAF aircraft capable of speeds above three hundred miles an hour, the Hurricane had been the mainstay of the defence against German bombers in the Battle of Britain.

Accordingly there was this entry in 71 Squadron Official Diary for 7 November 1940:-

"Everything good happened today. Nine Hurricanes arrived, delivered by No. 85 Squadron. The Commanding Officer, Squadron Leader Townsend — a most attractive fellow — was leading his squadron into battle daily despite one foot in plaster-cast, having had several toes taken off by a cannon shell. Further eight pilots posted from No. 5 Flying Training School, Sealand arrived".

That "most attractive fellow" leading the Hurricane flight, the first fit-for-combat aircraft to be delivered to an American unit in World War II, was Peter Townsend, who was later to become well known as equerry to King George VI and as a frequent escort of Princess Margaret.

Newly arrived No. 71 Squadron pilots, all Pilot Officers, were Pete Peterson; 'Indian', Jim Moore and Gus Daymond; Charles Edward Bateman of Stephentown,

New York; and Californians Stanley Michael Kolendorski from Bell; Byron Kennerly from Pasadena; James Leland McGinnis from Hollywood and Edwin Ezell Orbison from Sacramento.

These men were added to a nucleus of earlier arrivals and were soon followed by more volunteers. The young Eagles were eager to fly and fight, and their high spirits showed.

Meanwhile, Colonel Sweeny announced at a news conference in Ottawa, that the formation of the Eagle Squadron had moved forward so well that he was retiring from his association with the unit.

"The men are in good hands, and they don't need me any more", the old soldier said. In London, Charles Sweeny arranged for his brother, Bobby, to become Adjutant of 71 Squadron.

Walter Churchill proved to be the ideal Commander to launch 71 Squadron on its way to excellence. *"He was a strict disciplinarian, but we came to love that man"*, said Pilot Officer Vic Bono of him later.

When Bill Taylor returned to 71 Squadron from Spitfire indoctrination, he went to London to remind the Air Ministry of the assurances that had been made to him.

The RAF there-upon promoted Churchill to Wing Commander, in control of three fighter squadrons. In January 1941, Taylor at last was in command of 71, an American leading American pilots in a British unit.

Bobby Sweeny had found at 71 Squadron a job-overlap situation not unlike that involving Taylor and Churchill. British Flying Officer F. H. Tann had become Squadron Adjutant, the post that Charles Sweeny had lined up for his brother Robert. Bobby Sweeny, in effect, became the assistant adjutant, an office that did not exist in the RAF table of organisation. *"I was educated at Oxford, and Air Marshal Sholto Douglas, Chief of Fighter Command, thought I could keep the Squadron in line so they wouldn't break too many RAF regulations"*, Bobby Sweeny recalled. *"What a hope! On their night off, these fellows were like a pack of Sioux Indians on the warpath. I must have visited most of them in most of the jails of England bailing and talking them out, wherever we were stationed, mostly in Lincolnshire"*.

"I left the Squadron late in 1940. After a little flight training, as I already had 150 hours, I transferred to 247 Squadron RAF Coastal Command, flying B-24 Liberators". He later was awarded the DFC for bringing home a badly damaged aircraft after a particularly difficult encounter with a German U-Boat.

Further extracts from the Operations Record Book, chronicles the programme of events of No. 71 squadron whilst at Church Fenton:

Date	
1.10.40	Flight Lieutenant G. A. Brown arrived from RAF Station, Kenley.
2.10.40	Pilot Officer F. N. B. Bennett arrived (Engineer Officer)
4.10.40	Squadron Leader W. E. G. Taylor and Pilot Officer R. V. Sweeny arrived. Squadron Leader Taylor had been released to the RAF from the Fleet Air

Arm having seen service on HMS Glorious in Norway and First Lieutenant R. V. Sweeny had completed an administration course at Loughborough.

7.10.40 Parade for Group Captain C. Sweeny.

12.10.40 Pilot Officer P. H. Leckrone arrived. Posted from No. 616 Squadron, Kirton-in-Lindsay. His squadron were very reluctant to release him.

14.10.40 Pilot Officer L. E. Allen arrived. First pilot of the new lot that were training at Sealand to finish his course, having been pushed through in record time.

21.10.40 Flight Lieutenant R. C. Wilkinson, DFM (and Bar) arrived from No. 3 Squadron, Castletown, Scotland.

23.10.40 Pilot Officer A. G. Donahue taken back into his own Squadron, unhappy with us due to complete lack of any aeroplanes at all, barring one unserviceable Miles Magister. At this point the entire personnel were "completely browned-off" because of the enormous lack of interest in equipping the Squadron with any aircraft.

28.10.40 Pilot Officer Leckrone overshot the field this morning in the manner in which the Brewster float on landing, turned the aircraft over on its back at a good speed and gave himself a bad crack on his head.

29.10.40 Pilot Officer Leckrone admitted to hospital with slight concussion. During this period the remaining two Brewsters were flown continuously and Squadron Leader W. M. Churchill, DSO, DFC drew up a report on their performance with Gibson, the Brewster test pilot, who had visited the Station, and sent it back to his company in America.

1.11.40 Big send-off to No. 306 Squadron (Polish) who were going on operational service.

3.11.40 Received signal that the Squadron was to be re-equipped with Hurricanes. Squadron Leader Churchill had an alarming experience in one of the Brewsters — one of the inspection panels in the tail came open while he was flying and the rush of air on the aileron control made it almost impossible to control the plane. After considerable difficulty he landed at Linton-on-Ouse aerodrome close by and returned the plane after finding the trouble. Another send-off to the Poles, who definitely promise to leave tomorrow.

6.11.40 Eric Kennington, the artist, arrived to do portraits of Squadron Leader Churchill DSO, DFC and Flight Lieutenant R. C. Wilkinson DFM having already done Group Captain C. Sweeny and Squadron Leader W. E. G. Taylor.

15.11.40 Station was terribly pleased to hear that the first V.C. given to a Fighter Pilot was awarded by the King to Flight Lieutenant Nicolson (see separate chapter) who had been stationed at Church Fenton some time prior to our arrival.

16.11.40 R/T practice to-day. The pilots went to 30,000 feet. Flight Lieutenant G. A. Brown was kind enough to say that he thought this squadron would be very good once they were trained.

19.11.40 Flight Lieutenant Hector Bolitho visiting to get material for an article for the "Saturday Evening Post", a weekly paper with a circulation of over 20,000,000 in America.

21.11.40 "Thanks Giving Day". A much appreciated effort was made to give this squadron a Thanks Giving Luncheon. The Squadron was posted to Kirton-in-Lindsey. We were sorry to leave as everyone had been so kind!!

Charles Sweeny concludes;

"In May 1986 — very approximately in the centre of London, in Grosvenor Square, opposite the Roosevelt Memorial and in front of the American Embassy, a very impressive memorial was unveiled that identified forever the brave deeds of the Eagle Squadrons. The unveiling took place in the presence of the Prime Minister, Mrs Margaret Thatcher, the US Ambassador Charles Price, and a distinguished gathering composed of their friends and the heads of the Army Services of Great Britain and the United States.
This memorial, set out in letters of gold, the names of all the brave Eagles whose deeds and courage will not be forgotten".

Americans 'Red' Tobin, 'Shorty' Keough and 'Andy' Mamedoff — all killed in action later.
(Photo: Courtesy of Chaz Bowyer)

71 EAGLE SQUADRON, CHURCH FENTON NOVEMBER 1940
Standing L to R: P/O Kolendorski, P/O Daymond, P/O Keough, P/O Mamedoff, P/O Allen, P/O Kennerly, P/O Peterson, P/O Orbison, P/O McGinnis.
Seated L to R: F/O Bennett (E.O.). F/L Brown, S/L Churchill, F/L Wilkinson, P/O Tann (ADJ).
(Photo: Courtesy of James A. Gray — Eagle Squadron Association)

6

CHEQUER BOARDS IN THE SKY

A SHORT HISTORY OF THE POLISH AIR FORCE IN BRITAIN

'THE CHEQUER-BOARD'. (historical)

The origin of the Chequer-board emblem carried on aircraft of the Polish Air Force (PAF) is kindly supplied by the Polish Institute and Sikorski Museum:-

Lt. Stefan Stec, whilst still serving with the Austrians, had the chequer-board insignia painted on his aircraft. During the Polish defence of Lwow against the Ukrainians in November 1918, Lt. Stec flew to Warsaw, and the insignia was so liked by Air Force Lt.-Colonel Hipolit Lossowski, the Chief of the Aeronautical Department in the Ministry of War, that he issued a General Staff Order (No. 38 of 1.12.1918) instructing that it be used in future as the official PAF emblem.

'THEY WANTED TO GO ON FIGHTING'

The faith of her airmen enabled the Polish Air Force to be reborn in Britain. The infamous and totally unprovoked attack on the Polish nation on 1 September 1939, without declaration or previous intimation, heralded the continuation of the Nazi 'blitzkrieg' against undefended cities and innocent civilians.

The Polish Air Force fought valiantly and skillfully against the Luftwaffe armadas and inflicted heavy losses, but were no match for the numerically superior and better equipped formations. The added threat to the Polish forces already desperate situation came with the invasion of the Eastern borders by the Russians, compelling them to cross the borders into Hungary and Rumania. From internment camps, the beleaguered Polish Air Force quickly organised escape routes to France and England. Poland's General Sikorski negotiated with the British Government the first of three agreements which allowed for the operation of a separate Polish Air Force within the framework of the Royal Air Force. The perilous position of Britain after the fall of France accelerated the formation of efficient and much-needed Polish operational squadrons, their skill and bravery to become legendary.

THE RESIDENCY OF THE POLISH SQUADRONS AT CHURCH FENTON

NO. 307 (Polish) — "The Eagle Owls of Lwow
Squadron Anniversary — 10th September
Squadron designation — EW
Squadron Commanders at Church Fenton.
Wing Commander G. K. Ranoszek (4.5.44 tó 20.12.44)
Wing Commander S. J. Andrzejewski (21.12.44 to 2.3.46)

The squadron was made up from original members of the Vilno and Lwow Air Regiment, and was formed at Blackpool on 24 August 1940, and moved to Kirton -in-Lindsey on 5 September 1940 to receive Defiants. On 7 November they left for Jurby, Isle of Man, and reached operational status on 3 December, engaged on convoy escort by night and daylight duty in bad weather. By 27 January 1941 the unit had moved on to Squires Gate and commenced patrols at night in defence of Merseyside and the Midlands. On 26 March 1941 they moved on to Colerne, near Bath, moving on again to Exeter on 14 August, where the Squadron was resident for two years. During August they were rearmed with Beaufighters Mk. V, and most of the defensive victories were using these aircraft.

Most notable were four German bombers destroyed on the night of 3/4 May and a further three on 4/5 of August 1942. Mosquito Mk. 11F with A1 (Mk 5) replaced the Beaufighters by December 1942, and further changes of base followed during which time the squadron flew intruder and Ranger operations over Germany, France and the Bay of Biscay patrols. from the Shetlands they operated over Norway and Denmark, armed with Mosquito's Mk VI and XII.

On 24 May 1944, the squadron arrived at Church Fenton to re-equip with Mosquito Mk. XXX fitted with Mk. 10 'Monica' radar. During this period 307 operated as a night-fighter unit for defence and intruder operations over Norway, Denmark, Germany, Holland and Belgium. On 14 January 1945, they moved on to Castle Camps, near Cambridge, to be deployed in high-level bomber support duties over Germany. Besides the many ground targets that they attacked over enemy territory, the squadron destroyed in the air — mostly at night — 33 German aircraft, nine probably destroyed and eighteen damaged.

The squadrons own song — "The Eagle Owls of Lwow" — was composed by their own members, and it was through the initiative of the squadron that the concept of the Polish shrine in Liverpool Catholic Cathedral was embodied.

The squadron was disbanded at Coltishall on 2 January 1947.

No. 306 (City of Torun) Squadron
Squadron Anniversary 4th September — designation UZ

The squadron was originally formed in France in the Spring of 1940, mostly from the members of the Torun Air Regiment and a small number of the Poznan Air Regiment. On the collapse of France it was transferred to Britain and made its way to Church Fenton to re-form and re-equip. Polish aircrew and supporting staff began to arrive during the period 6th June to 4 September 1940. This was to form the nucleus of the squadron, commanded by Squadron Leader T. M. Rolski. He recently described in his own words those epic days:-

"The squadron spent this period usefully and were accepted into the Yorkshire community with kindness and hospitality. We trained on Battle and Master aircraft — later Hurricanes Mk. 1 — and carried out night-flying exercises aided by lights and navigational beacons which emitted a red recognition signal in Morse.

During the first months there were language difficulties and organisational matters, but we were well supported by R.A.F. liaison officers. The Commanding Officer was Squadron Leader Douglas Scott, and in the flights were Flight Lieutenants Kennard and Tennant. On the Polish side there was myself as Commanding Officer and Flight Lieutenant F. Skiba; we had all facilities for training in flying, and living. All our pilots were eventually trained on the Hurricane Mk. 1 without accident — excepting one — who approached to land but forgot to lower his undercarriage. He made a successful belly landing, but was later posted to another unit. We used the Link trainer for instruction on night-flying. In the period 1941/1942 our squadron had the smallest number of flying accidents in Fighter Command and received a Commendation from the Air Council"

On 5 November 1940, No. 306 was classified as fully operational and departed from Church Fenton to take up duty with No. 9 Group at Ternhill, and later fought with distinction in the defence of the Southern counties. Later in the war the Squadron was responsible for destroying 60 of Hitlers Wunderwaffe (flying bomb) and seventy enemy aircraft, including twenty-nine damaged and sixteen probables.

The squadron was eventually disbanded on 6 January 1947 at Coltishall.

Editors note : *Squadron Leader T.M. Rolski later became Group Captain and a recognised authority and tactician in air combat. His eventual military decorations included Polands highest award, the Virtuti Militari, the Polish Military Cross of Valour (with two Bars) and the British D.S.O.*
(Photo: Courtesy Squadron Leader T.M. Rolski)

Wing Commander S. Andrzejewski Commanding Officer of No. 307 (Polish) Squadron (right) with Group Captain Bajan, Liaison Officer.
Military decorations: Virtuti Militari, Krzyz Walecznyck and British Air Force Cross (A.F.C.)
(Photo: Courtesy Wing Commander S. Andrzejewski)

Wing Commander G. K. Ranoszek in his Mosquito
(Photo: Courtesy Wing Commander G.K. Ranoszek)

Pilot Officer J. Zulikowski was one of the youngest of the pilots of No. 306 (Polish) Squadron when it was formed at Church Fenton in 1940. In May 1943 with the rank of Squadron Leader, he commanded No. 308 (Polish) Squadron here during its period of residency. His military decorations include the Virtuti Militari, Cross of Valour — with two Bars, and the British D.F.C. He was also the last wartime Commanding Officer of No. 306 Squadron.
(Photo: Courtesy Pilot Officer J. Zulikowski)

7

No. 85 SQUADRON

A BRIEF HISTORY

Badge: *On an ogress a vioded hexagon.*
Motto: *Noctu diuque venamur*
(We hunt by day and night)

Few squadrons have packed so much into its history as No. 85. Formed at Salisbury Plain on 1 August, 1917, it was in action in France by the following May under the command of Major "Billy" Bishop VC, a Canadian, and one of the most famous of WWI aces. He was succeeded by Major "Mick" Mannock, VC., the one-eyed victor of 73 aerial combats whilst he was on the Squadron, and the acknowledged leading British air-fighter of the war. The white hexagon symbol was first seen during this period, and was thought to have originated from part of the coat-of-arms of a French town near the Squadron base. This method of air to air identification painted on the aircraft, using geometric symbols, became the practise for all fighter squadrons. The white hexagon is perpetuated in the Squadron crest. The squadron was disbanded in 1919 in line with the world-wide disarmament arrangement between the combatant nations.

However, in June 1938, the Squadron was reformed at Debden — with Gloster Gladiators — a four-gun, highly manoeuverable biplane, destined to give excellent service in many campaigns. In September of that year conversion to Hurricanes took place, just one day before the war was declared, and a year later the unit moved to France as part of the Air Component of the British Expeditionary Force.

The first German aircraft destroyed was a Heinkel III in November, 1939, but activity was fairly quiet until the main enemy attack in May, 1940, when the Squadron was in action continuously, suffering a severe mauling in the process. By the end of the month the tattered remnants were hastily evacuated, having inflicted considerable damage to the enemy; they could claim 90 enemy aircraft destroyed and many damaged. Seventeen of the forty pilots of 85 Squadron were casualties, but in recognition of their fine effort nine decorations were awarded.

At this point of the story Group Captain (then Squadron Leader) Peter

Townsend assumed command of the depleted — but not demoralised squadron.

On 23 May, 1940, he arrived at Debden, Essex, with instruction to reactivate the Squadron to its former fighting efficiency. So swift was the German advance in France that much of the Squadrons' equipment — and some ground crew — had to be left behind, much to the chagrin of the pilots. The sadly decimated Squadron made its way back in fragmented fashion to Debden by all manner of transportation — by air (only four Hurricanes made it to fight another day) — some by trainer aircraft, some by train and car. He awaited their arrival with a welcoming smile, and some apprehension. 85 had been a crack squadron, with a great tradition of skill, gallantry and devotion to duty. They had recently lost two Squadron Commanders — 'Doggy' Oliver, who was badly wounded — and Michael Peacock, killed in action. He was to be the third CO and was anxious to discover what their reaction would be to their new leader:

Happily, they became good friends. One of his main problems was pilot replacement as a dozen new pilots now on strength were but youths in their late 'teens and early twenties. Their flying experience was basic with very little to claim on the powerful and speedy Hurricane.

However, the more experienced of them nurtured the new fledglings in the skies over Debden, drilling them in the disciplines and tactics of air combat, and by the end of June the squadron was operational once more. The main preoccupation was the dawn-to-dusk patrols to protect the constantly menaced coastal convoys from enemy bombers. The more experienced pilots participated in night-flying operations but the Hurricanes lack of RADAR equipment reduced the chances of success. Day operations took place from airfields in the London area during the Battle of Britain and 85 took part in most of the bigger air battles — including one of its own on 18 August in the Canterbury area. Two hundred Hun aircraft were encountered, and although the odds were 15:1 against them, ten enemy aircraft were destroyed and a further ten damaged — for the loss of one Hurricane.

The intensity of the air battles increased as the month progressed, and 1 September 1940 was a bad day for Fighter Command — disastrous for 85 Squadron. Whilst patrolling over the Kenley area they encountered a numerically superior enemy formation and during the ensuing air battle they suffered heavy casualties:-

** Flight Commander Patrick Wood-Scawen was shot down and tragically died when his parachute failed to open.

** Flight Sergeant Booth — shot down and suffered severe back injuries when his parachute failed to open properly.

** Flight Sergeant Allard — landed at Lympne with engine trouble after combat, his aircraft being destroyed when the aerodrome was bombed.

** Flying Officer Gowers — shot down and suffered severe burns and wounds to feet and hands.

** Flight Sergeant Ellis — shot down and killed.

** Pilot Officer Lewis — suffered damage by Bf. 109 over Biggin Hill and landed wheels-up at base — pilot unhurt.

On the 3 September the squadron was withdrawn from the front line and sent to Church Fenton for a rest — and to re-equip, this time led by Flight Sergeant Sammy Allard, who was the Squadrons best pilot. At this time Squadron Leader Townsend lay helpless in his hospital bed recovering from an operation to remove a German cannon shell from his left foot, thus again rendering the squadron without leadership. To his great delight he arrived at Church Fenton shortly afterward and made for the dispersal point, with his foot encased in plaster. The foot wound prevented him from walking except with the aid of a stick but he could still fly. Willing hands assisted him into the cockpit of a Hurricane and he then proceeded to put it through its paces. That day the Station Doctor cleared him as fully fit to resume command of the squadron.

During this short rest period at Church Fenton, 85 received orders to convert to a night-fighter squadron and moved to Kirton-in-Lindsey, Lincolnshire. During the next five months the unit moved to Gravesend and later Debden, and the squadron performed its night-fighter role in Hurricanes. In March 1941 the pilots returned in groups of three at a time for a short conversion course on 'twins' — Blenheims as it happened. Squadron Leader Townsend was delighted to learn that he was to be 'converted' by no less a person than Group Captain George Stainforth, who in the Schneider Cup Vickers-Supermarine S6 — forerunner of the Spitfire — had in the early thirties exceeded 400mph. Stainforth was one of his boyhood heroes, and in the hands of such a master, despite his instinctive dislike of twins, he was quickly converted; after twenty minutes solo on a Blenheim he was heading south again for Debden. It should be explained that the reason for twin course was in preparation for the new American, tricycle undercarriage, twin-engined Havoc equipped with the new A.1 (aircraft interception) RADAR, which would replace the Defiant and Hurricane for night-fighting.

Initially over England and then after 'D' Day, 85 was engaged on night intruder operations over the Continent and they added a further 134 enemy aircraft to its tally. During this period Wing Commander John (Catseyes) Cunningham was in command.

When the war ended the squadron was based in Norfolk and had accounted for 278 enemy aircraft destroyed, and a further 120 probables — or damaged. In total 97 decorations had been awarded for valour.

Since the war the squadron strength progressed from Mosquito's to Mèteors, and on 23 September, 1957, they returned to Church Fenton after an absence of 17 years. On 19 December, 1975, the squadron was disbanded at West Raynham and assumed the role of a Bloodhound Missile unit.

Hurricane 1 aircraft of 85 Squadron on patrol, flying in standard RAF battle formation
(Photo: Courtesy Imperial War Museum)

No. 85 Squadron in the night fighter role. Picture shows Squadron Group relaxing with Group Captain Peter Townsend and mascot "Kim". Squadron tally of enemy aircraft destroyed is recorded on German aircraft propeller recently shot down.
(Photo:: Courtesy Imperial War Museum)

8

N0. 54 OPERATIONAL TRAINING UNIT

On 10 August 1940 it was decreed that the sector station of Church Fenton be transferred from the operational control of No. 13 Group to No. 12 Group, the Air Officer Commanding being Air-Vice Marshal R. E. Saul MC DFC. With this transfer came a complete change of operational status — from defensive to training — to be known as No. 54 Operational Training Unit, the modus operandi being the training of pilots and Radio Observers (RO) for RADAR assisted night fighting, a technique comparatively new at this time, with many hard lessons to be learned. (See also notes in chapter entitled 'Interregnum')

With the departure of the last operational squadrons for other defensive duties the embryo No. 54 OTU. began to take shape, having the distinction of being the first of several Training Units instituted by Fighter Command to meet the increasing demand for well-trained night-fighter crews. Initially the Unit was under the command of Wing Commander Hogan DFC, to be replaced in December by Wing Commander R. L. C. Atcherley DFC (Batchy) who, with his twin brother David had their family home in nearby York. Both these officers were destined to attain Air Rank later and were both very popular 'characters'. The Chief Ground Instructor for part of this period was Squadron Leader James Nicolson VC who was recovering from burns and injuries sustained in action recently, which led to the award of the Victoria Cross — the only one to be awarded to a pilot of Fighter Command in the war. (see Chapter 14).

Initially the training aircraft available were both single and twin-engined types — Defiants, Blenheims Mk. 1 (short-nose), Oxfords and latterly Beaufighters, but from June 1941 onwards only the twin-engined types were used, the single engined committment being allocated to No. 60 OTU.

No. 54 soon gained the reputation of being the crack night-fighter training unit in the country, but not, unfortunately, without heavy cost in trainee aircrew. This could be attributed to several reasons:

(a) It is true to say that most OTU's had a fairly high casualty rate due to the inexperience of the trainees. If the mortality rate was higher at 54 OTU than at some other Bomber or Day Fighter OTU's this was not a criticism of the standards there but rather a reflection of the times and circumstances. Although not generally appreciated at the time, the aircrew were being trained as the first all-weather pilots — twenty four hours a day, every day, despite prevailing weather conditions — so they flew in weather that could have grounded some other training establishments, a statement perhaps refuted by the Bomber OTU's.

(b) Luftwaffe night intruders (the Fernnachtjager) normally operated converted Junkers JU.88 bombers painted all-black with the front turret replaced by a plain nose, with radar. As well as inflicting losses on our returning bomber stragglers they also directed their attentions to the Operational Training Units, resulting in heavy losses to the crews under training, as will be gleaned from the following examples:

The night of 24 April 1941 was to prove a particularly busy and eventful one for No. 54 at Church Fenton. The base was a hive of activity when Leutnant Heinz Völker paid a visit. His first victim was Blenheim L. 7132 flown by Flight Lieutenant Bill Dennison, flying solo. He was attacked and the petrol tank exploded, the port wing and engine bursting into flames. Dennison managed to bale out and landed safely. Völker then attacked Sgt. Spurgins Blenheim L.1478 which was in circuit, badly damaging it before it finally evaded him.

Sitting on the runway waiting for the signal to take off was Sgt. Attenborough in Blenheim L. 1320, which was bombed by Völker causing severe damage.

No. 54 and Church Fenton were the targets for Leutnant Pfeiffer two days later. This time a warning was given and visual signals were made instructing pupils to extinguish their navigation lights and orbit a beacon until the danger had passed, but for some reason Sgt. Crozier and Sgt. Bell (the turret gunner) in Defiant N. 1568, failed to see, or heed, the warning. At 1000 feet Pfeiffer attacked the illuminated aircraft and Sgt. Bell returned the fire, Sgt. Crozier taking violent evasive action to escape the cannon shells from astern. By this time both aircraft were flying at almost zero feet and in the darkness trees proved difficult to see as they loomed up quickly. The Defiant sliced its way through the treetops, finally crashing completely wrecked near Thorpe Arch Church, killing both crew.

On the 18 July, Sgt. Neighbour, a New Zealander, was making his first night solo flight and it was known that there were intruders in the region of Church Fenton. He was ordered to orbit a flashing beacon until the danger had passed but as the Blenheim circled it began to lose height, unnoticed by the inexperienced pilot, who was flying on instruments alone. Eventually the aircraft flew into the ground, the pilot perishing in the crash.

In 1941, the station was intruded upon thirteen times, and on five of these occasions bombs were dropped. As a counter-measure to these attacks defensive tactics had to be implemented, and in February 1941 No. 46 Squadron and their Hurricanes, under the command of Squadron Leader Rabagliati, moved into nearby satellite station of Sherburn-in-Elmet. They carried out many operations against the

Fernnachtjager before leaving in May, and were replaced by Nos. 485 RNZAF (Spitfires)... 129 (Spitfires)...601 (Airacobras) and 19 (Spitfires) Squadrons to provide full Sector protection. They operated from the satellites — Leconfield, Acaster Malbis and Hutton Cranswick, and on many occasions from Church Fenton itself.

No. 54 OTU — AND THE BRISTOL BLENHEIM

by Jeremy Howard-Williams

The Blenheim was not a difficult aeroplane to fly, but it was a good deal heavier and faster than the rather staid Anson, which had been my training aircraft until now and which was of a forgiving nature. You had to control a Blenheim, an Anson would fly itself. An indication of the accident rate was given, had we but known it, by our reception at No. 54 OTU. The Sergeants' and Officers' Messes were always full to capacity on the arrival of a course. It was quite usual for aircrew to be billeted out or even put under canvas and, on their protest, to be assured that there would be room in the Mess within a week. What they were not told was that these vacancies were caused either by Sir Isaac Newton taking personal control of an aircraft in bad weather, or else by the close attentions of the Luftwaffe. I arrived on 21 October 1941, not knowing anything about this and also not knowing that Hitler himself had issued an order the week before, revoking permission for the Luftwaffe to mount intruder operations. British aircraft, he said, must be destroyed over Germany where the public could see them coming down, and not over England — even though the scoring rate of the latter was better from the German point of view.

"Jeremy Howard-Williams and Blenheim R. 3760 October 194
(Photo: Courtesy Jeremy Howard-Williams)

12 November 1941

First solo on the Blenheim. I climb into the cockpit, check all round, open the cowling gills, see that the hydraulic lever is switched to work the udercarriage and flaps (and not the gun turret), that the undercarriage lever is down, the fuel selected to tanks with adequate supplies, the propellers in fine pitch, the fuel mixture control is on rich and the flaps up. I switch on and stick my head out of the port window to call to the groundcrew who are operating the external battery — the trolley acc as it is known.

'Contact port!' The cry is repeated. I press the tit and the engine roars into life. I adjust the throttle and then my thumb goes up and I point to the starboard engine. The mechanic imitates the signal to show that all is well and the other engine starts. The trolley acc is unplugged and wheeled away and I glance over the various instruments while I am waiting for the engines to warm up.

A second check on the hydraulics selector lever and I then put down my hand to exercise the flaps, taking care not to confuse levers with the undercarriage; the two controls are side by side and, of all stupid arrangements, have almost identical levers. I look down to ensure that I have the correct one and then operate the flaps successfully. I press button 'A' on the R/T control box to get flying-control frequency. As soon as the set is warmed up I run each engine in turn up to zero boost and check for excessive drop in revs as I switch each magneto off in turn; we are allowed a 50 r.p.m. drop, providing the engine is not too rough.

Having found the revs OK, with oil pressures and temperatures at correct figures, I stretch over my left shoulder and put each propeller into coarse pitch and watch the drop in revolutions. When I have returned the levers to the fine position I am ready to check the brake pressure and wave away the chocks from the wheels.

'Hello 62'. Flying control answer my call immediately; they have been warned that this is a first solo. 'Marker here. OK to taxi to east/west runway. Listening out'.

I taxi carefully out and stop on the perimeter track just short of the beginning of the runway.

'PMCRAFTS,' I mutter to myself, quoting the mnemonic for the cockpit check before take-off.

P. Petrol on and enough. Pitot head heater on.
M. Mixture rich.
C. Cylinder-head temperatures OK.
R. Retractibles: Hydraulics off turret and on wheels and flaps.
A. Airscrew fully fine; throttle nut tight.
F. Flaps up.
T. Trim; rudder neutral, elevator slightly nose down.
S. Sperry gyro set to compass heading.

'Good. Now close the engine gills'. I glance up and see that nothing else is coming in to land.

'Hello Marker, Circus 62 calling. Permission to scramble over.'

'Hello 62, Marker here. OK. to scramble. Listening out'.

I taxi forward to the runway. For the first hundred yards or so it is grass, as the tarmac strip has not been extended to the perimeter track.

'I'd better wait until I get properly on the runway before I open up,' I think. Then I change my mind. 'No. I'll start now, I won't be going fast by the time the tarmac starts'.

I open the throttles and the Blenheim immediately leaps forward.

'Too fast for this grass, I think, and start to throttle back. 'Too late. I'll be going fast anyway, so I'd better rush it.' I push the throttles fully open and hang on for dear life. The Blenheim leaps about over the uneven surface and I am a little scared. The beginning of the tarmac runway soon arrives and the aircraft bounces sharply. It swerves to the right and the straightens up as I apply left rudder.

My starboard wheel is now off the runway and bouncing along the grass on the right. The rest of the aeroplane is on the tarmac surface, but I decide to press on with my take-off. The Blenheim is bucking about and gathering speed rapidly. I manage to hold it straight and notice with dismay a crowd of workmen looming up in front. Shortly before we reach them at the end of the runway the aeroplane takes to the air. This it does with no conscious effort of mine and we swing crazily away, climbing slowly.

My next actions are completely automatic. Wheels up. Throttle back to cruising boost. Propellers into coarse pitch. The aircraft proceeds to fly me round the sky for the next five minutes, during which time I am content just to sit at the controls.

'OK Jesus, I've got her.' I come out of my semi-coma and grin to myself at the old joke about the Good Lord taking control for the first five minutes of a first solo, which I can now better understand having been a perfect illustration of it. I resume control and discover that I am sweating freely. The take-off has frightened me considerably and I am glad to have got safely out of it. The Blenheim gives the impression of having a good deal more metal about it than any aeroplane I have yet flown. It is all sharp corners inside and already my knuckles are cut and bruised from operating the inconveniently placed tits. It passes through my mind that a crash in one of these could have unpleasant consequences. (I am to try this out at first hand before the end of the course, only to prove that the Blenheim is, fortunately, not a killer every time.)

After an hour, during which time I managed to cool off a little, I wipe myself down thoroughly so as not to reveal my state to the groundcrew (as if they could be fooled), return to the airfield and make an uneventful landing. After learning to fly the Blenheim by day, we had then to move into D Squadron to fly it by night and to team up with a radio operator before starting to find out all about A1. The electronic equipment, which was later christened radar by the Americans _ from the initial letters of Radio Direction And Range — was at that time known publicly as radiolocation in England, but the fact that we had radar in aircraft had not been released.

The equipment in squadron use was A1 Mk. IV, which transmitted radio pulses all round the fighter aircraft, thus 'flood-lighting' the surrounding airspace. This restricted its maximum range to the height of the fighter, because the ground beneath sent back a massive echo which swamped the set from that range outwards. Any aircraft flying closer than the fighter's height, therefore, would send back an echo which could be seen in the clear part of the cathode-ray tubes. It was then the radio operator's job to interpret the signals he saw and to instruct the pilot what to do: turn right or left, climb or dive, speed up or slow down.

Before you get the impression that this is easy, let me remind you that the A1 sets of those days were far from perfect and were constantly breaking down or giving false readings. Even if the sets were working well and the target nicely within range, the RO saw indications on his CRTs telling him only roughly how many degrees the target was off the centreline of the fighter, both in the azimuth and elevation planes. He had no direct indication of which way the target was heading or whether it was in the act of turning, accelerating or slowing down: all this information he had to infer from change of angle and change of range. Even in those days things happened quickly enough to make it vital to give the order where to go first, followed by how quickly to do the manoeuvre and then information on the target's movements, thus:

'Left, hard left. Tighten the turn, he's crossing starboard to port range about two miles. Ease the turn. Steady, he's 5º port and 5º up. Range still two miles. Port again gently. Steady. Dead ahead now, still slightly up, throttle back a bit, range coming down rather fast.'

A1 Mk. IV DISPLAY

In A1 Mk. IV, a dipole aerial fixed in the nose of the aircraft, broadcast radio waves on 11/2m, in pulses, to "floodlight" the sky all round the fighter aircraft. Receiving dipoles were positioned above and below either wing, so that they could measure the difference in the return time of the radar pulse as it echoed back from any large object which was covered by the radiations. These time differences (only microsec) were suitably displayed on two cathode-ray tubes in the navigator's cockpit, and showed the angle off in both elevation and azimuth.

Because the transmitter floodlighted the sky, it follows that the ground beneath the fighter received as many pulses as did the sky in front. This meant that a massive return echo was received from the ground, at a range equal to the fighter aircraft's height, which effectively drowned out any smaller target such as another aircraft, if it were beyond that range. Thus, a fighter equipped with A1 Mk. IV which was flying at just over 5,000 ft. altitude had a maximum radar range of one mile, while at 21,000 ft. this went up to four miles. Normal minimum range was about 800 — 1000 ft.

No. 54 O.T.U. (the Blenheim — Oxford incident).

Another enduring partnership that was initiated at Church Fenton with No. 54 OTU was between Sergeant Pilot Harry White and Radio Observer (U/T) Michael Allen, who 'crewed up' on 4 August 1941. They were both 18 years of age and after 47 years are still good friends, and both won the DFC and two Bars in October 1944.

An incident that happened under training is narrated by Sergeant Pilot Harry White (now Air Commodore) with further decorations CBE and AFC.

"I was involved in a mid-air collision between a Blenheim and an Oxford on the 20 July 1941. The wind had been changing direction most of the afternoon leading to a frequent change of runway for take-off and landing. I was taking off in an Oxford in one direction and as I approached the intersection I saw a Blenheim bearing down on me from the right obviously intending to land on the intersecting runway. Although only just airborne I pulled up the nose of the Oxford in an attempt to climb over the Blenheim approaching on my starboard side. The Blenheim pilot responded at precisely the same time and I saw his nose go up in the air at the same time as mine. Both aircraft collided at about 50 feet above ground. His aircraft churned my starboard wing and hit the starboard engine which stopped. This inverted his aeroplane and he landed upside down about two or three hundred yards down the airfield.

With no starboard wing or engine my port wing came up and over and wrote itself off on the ground. What was left of the aircraft not only completed the 360 degree roll but also turned through 180 degrees in direction and slithered backwards before coming to a halt. Although the participants were shaken none were really injured.

On the training at Church Fenton at that time all I can say, now, is that the blind were leading the blind. No-one knew much about night-flying and even less about asymmetric flying. A lot of crews were killed during my three months at Church Fenton, a high proportion at night, and several because of engine failure.

Those who got there were almost lucky to survive. The Pennines too collected their fair share of wreckage as people improperly let down through cloud to try to establish their position"

As a crew they left Church Fenton, Michael Allen having qualified as Radio Observer and promoted to Sergeant, and posted to No. 29 Squadron on Beaufighters at West Malling. They completed one tour in Fighter Command and two in Bomber Command, completing 106 sorties and claiming twelve enemy aircraft destroyed and four damaged. They went their separate ways in November 1945, Harry White stayed in the RAF and Michael Allen into a career in Industry and Commerce.

Editors comments. *My thanks to Mr. Michael Allen for the photographs shown in this narrative and others not shown due to space restrictions.*

Boulton-Paul Defiant fighters being serviced
(Photo: Courtesy Imperial War Museum)

Remains of Oxford and Blenheim after mishap over runway — 20 July 1941.
(Photo: Courtesy Michael Allen)

No. 54 Operational Training Unit — Radio Observer Course — August 1941. Michael Allen is seated fifth from left. (Photo: Courtesy Michael Allen)

Bristol Beaufighter Mk. X
(Photo: Courtesy Imperial War Museum)

Hurricane preparing to take-off.
(Photo: Courtesy Imperial War Museum)

9

" WE PROWL BY NIGHT"

A narrative of 96 Squadron
by
Squadron Leader R. G. 'Tim' Woodman DSO DFC

96 Squadron was stationed at Church Fenton for only a short time — from 4 August 1943 until 3 September 1953. They arrived from Honiley (near Kenilworth) where they had spent almost six months as the night-fighter defence squadron for the south Midlands, and were in the process of converting from Beaufighters Mk VI to Mosquitos Mk XII, fitted with A.I Mk IV RADAR.

The squadron was originally formed during World War 1, on 28 September 1917, as a training unit at South Carlton, but had to be disbanded on 4 July 1918, as personnel were desperately needed for the reinforcement of operational units on the Western Front. It reformed again on 28 September 1918, at Wyton, and would have gone onto action as a ground-attack squadron, flying Salamanders, but for the Armistice and de-mobilisation.

At Crage in 1940 the squadron was again reformed on 18 December. They began operating as a flight of night-fighter Hurricanes, although they were normal day fighters with night camouflage and fitted with metal plates above the exhausts to eliminate dazzle at night, especially at full throttle on take-off. I flew these aircraft at Debden airfield but, in February 1941, the squadron received Defiant night-fighters,the type having suffered heavy losses at the time of Dunkirk, thus they were withdrawn to be used as night-fighters.

The squadron flew night patrols in defence of the North Midlands and Merseyside, having moved base to Wrexham. They commenced now a gradual conversion to Beaufighters Mk II, in May 1942, these aircraft having Rolls Royce Merlin engines with a poor single engine performance. Some squadrons, converting from defiants, had substantial losses during the process, numbers of crews being killed.

In September 1942, Beaufighters Mk VI began to replace the Mk II, and these had Bristol Hercules engines plus the dihedral tailplane, and were regarded as much safer aircraft.

The squadron moved on to Honiley on 20 October 1942, and I joined them in December after a six months spell seconded to the Bristol Aeroplane Company at Filton as a test pilot. German bombers were few and far between over the Midlands that winter and we spent a lot of time practicing interceptions on each other. During moonlight periods one could volunteer to operate out of Ford or Tangmere on RANGER missions over France, which was mainly train-busting.

I selected the Paris-Rouen main line and the lines connecting Laon-Rheims-Soissons-Compiegne, along which I managed to bust three locomotives one night. The Beaufighter had excellent forward vision for this type of operation - the rails glinting in the moonlight, looking ahead for a plume of steam, or the unfortunate fireman opening his firebox door. The French Resistance informed us that the midnight passenger train from Paris to Rouen was packed with German troops, and it was the practise for the officers to occupy the front and rear carriages. Although our primary objective was to incapacitate the locomotive, it was inevitable that these 'reserved' carriages were also given the treatment. These RANGER operations were excellent training for those of us who later volunteered for 100 Group Mosquito night-fighter operations over enemy territory.

I seemed to have had a love-hate relationship with the Beaufighter. At Filton whilst test flying the Mk VI with the new dihedral tailplane, I pulled out of a dive over the Mendips and the port elevator outer bearing bolts sheared, resulting in the elevator slamming into the rudder and locking it over. As I was contemplating the large hole that I was about to make in mother earth, the elevator broke away taking the rudder with it. Against all the odds, I managed to successfully land back at Filton. Within a month of joining 96 Squadron at Honiley, another incident occurred when flying a Beaufighter, which had been signed up by the ground crew, but not re-fuelled. My port engine cut, followed by the starboard as I was attempting an emergency landing. I crashed through the village of Ballsall Common, and was pulled out of the burning wreckage by my observer, Sergeant Witt. The aircraft was totally destroyed before anyone from Honiley reached the site. Wing Commander Burns (a South African) and the Squadron Commander was assured by local residents that there were no survivors from the crash. It was a shock when he turned round and saw me standing beside him.

A further hair-raising incident happened later whilst testing a Beaufighter that had been repaired after a crash and had suffered extensive damage. They had connected the elevator trim tabs the wrong way round, making for a very dicey take-off which nearly killed me.

Wing Commander Burns was replaced by Wing Commander E. D. Crew as Squadron Commander and he brought the squadron to Church Fenton. My memories of this busy station are not very strong due to our short stay there. Referring to my log book I see that I not only flew every day, but mostly twice a day too. On 22

August 1943, I flew four times, including a flight to Coltishall where we re-fuelled. We shortly afterwards took off again to take part in a MAHMOUD operation over the Zuyder-Zee, North West Germany and Denmark. After 96 Squadron crew taking part in MAHMOUD operations that night were F/O Robertson DFC and F/O Day DFC. After we were back over England I called him up but he, too, had had no success with German night-fighters. When I called him again he did not reply and I assumed he was on intercom. I landed at Church Fenton but Robbie failed to arrive. It transpired that he had collided with a USAF B17 Flying fortress over Norfolk, the American crew had been practising some clandestine operations they wanted to carry out. Robertson and Day and most of the American crew were killed.

At Church Fenton we patrolled for Patrington and Staythorpe GCI stations, giving them interception practice. On the evening of the 17 August, 1943, I took a party of 96 Squadron aircrew to the Patrington CGI station to meet the ground controllers, airmen and WAAF's, to let them know we were flesh and blood and not just call signs in the night skies. As it happened, we witnessed a small blitz on Hull, and as I stood outside watching the searchlights weaving to-and-fro, a German bomber was coned by a few, but escaped the concentration of light by diving steeply. I was amused to see a cat sitting alongside me, obviously fascinated by the searchlights. There was a whine of a falling bomb — if you can hear them coming they don't hit you, as they assured us from the trenches in the first World War — and it fell a short distance away with a colossal bang. I hared it to the nearest air raid shelter, and was amused to find that the cat had beaten me to it. Flight Lieutenant Lomas, of 604 Squadron based at Scorton, shot down a Do217 and claimed another as probable during this raid.

I managed to visit York one afternoon and must have partaken the brew at one or two 'locals' — but I cannot recall them. When more knowledgeable folk ask me what beer I drank my reply is "I think it was Smith's" — to which they answer "but which Smith's?"

The squadron left Church Fenton for Drem, near Edinburgh, on 3 September 1943, I left three days later, taking up a new Mosquito, where we continued our GCI practice at various locations and acting as top cover for a Navy carrier which was passing down the East coast of Scotland.

I left 96 Squadron at Drem in November, 1943, volunteering for operations with 100 Group. The squadron was disbanded as a night-fighter unit on 12 December 1944.

'Tim' Woodman

14.3.88

No. 96 Squadron members. S/L 'Tim' Woodman back row, second right.
(Photo: Courtesy of S/L Woodman)

Glossary of some Service Terms. (compiled by Squadron-Leader Woodman)

A1	Airborne Interception. Later more generally known as radar.
	Mk. IV. 11/2 metre. An all round coverage from fixed dipole aerials. Range limited to height due to ground returns on C.R.T.
	Mk. V. Adaptation of Mk. IV for pilot indication.
	Mk. VI. As Mk. V. but installed in single-seater fighters and Defiants.
	Mk. VII. 10 centimetres. A spirally rotating scanner in the nose of the aircraft gave a beamed transmission and thus could be used low down. Range 5-6 miles.
	Mk. VIII. Developed from Mk. VII. Range improved to 7-8 miles. Also incorporated beacon and IFF facilities.
	Mk. IX. A further 10 centimetre development with range up to 10 miles, and able to lock on to target. Other improvements incorporated.
	Mk. X. 10 centimetre but now with a fully rotating scanner.
BABS	Blind Approach Beacon System using radar.
Blip	Radar echo on the CRT.
CH.	British early warning radar, or Chain Home.
CHL.	The same but for low surveillance. CH Low.
CRT.	Cathode Ray Tube.
Eureka.	British beacon. 10 centimetre waveband.
Flensburg.	German electronic device enabling their night fighters to home on

to MONICA transmissions.

GCI Ground Control Interception.

Ground Return. Echoes on a CRT rrom the ground which obscured blips of target aircraft.

IFF Identification Friend or Foe.

Lichtenstein. German airborne radar for night fighters. 53 centimetres. A narrow beam of some 25 degrees width and maximum range of 2 miles.

Mahmoud. Night fightcr sorties over enemy territory to induce German night fighters to intercept.

Monica. Development of A1 Mk. IV for use as a tail warning device.

Mother. Beacons for R.A.F. airfields operating on 11/2 metres.

N.F.T. Night Flying test. Essential to check that all equipment was serviceable.

Radar. An American derived term which came to replace A1.

Ranger. Night fighter intruding into enemy territory, mainly for low level ground attack sorties.

Rhubarb. Low level offensive sweeps over enemy territory by daylight fighters.

SCR 720. From our inventions taken to the States by Sir Robert Watson-Watt the Americans developed their 10 centimetre airborne radar apparatus. The same as the British A1 Mk. X.

Turbinlite. A1-equipped aircraft (normally Havocs) which could shine a powerful searchlight on a target, thereby permitting escorting Hurricanes to shoot it down. Not very successful.

Whiting. The manouvre of a British night fighter when intercepted by the enemy night fighter. A turn through 360 degrees, more or less, to get on the enemy's tail.

10

No. 125 (NEWFOUNDLAND) SQUADRON

Motto: Nunquam domandi
(Never to be tamed)
by Warrant Officer (Pilot) Barry Radley

The squadron was originally formed in July 1918 at Old Sarum for light bomber duties, but was unfortunate to be disbanded in September of that year to provide supplementary personnel for other depleted squadrons. It was however given a new lease of life when reformed at Colerne on 16 June 1941 armed with Defiant Mk. I, supplemented by borrowed Hurricanes, to counter the German Baedecker raids against Bath in April/May. Operating as a night-fighter unit, the squadron patrolled an area covering South Wales and Western England, eventually converting to Beaufighter IIf. From February 1942 to September 1942 — supported by Defiants over this period. A further conversion took place in February 1944 when Mosquito Mk. XVII's began to arrive in strength, the squadron by this time operating from Valley and Hurn to cover convoys and bridgeheads, and sorties against V.I's during the invasion of Normandy. A further change of base to Middle Wallop found the squadron engaged in cross-country and low-level flying practice over the Atlantic and Welsh coast,and many operations over the Low Countries, frequently using Bradwell Bay as a forward base in order to gain range and increased time on patrol over Holland. A transfer to Coltishall in October 1944 saw the squadron mainly operating against the V.I.'s and Heinkel III V.I. bomb carriers, and the intruders who were infiltrating our returning bomber streams to try to attack them whilst they were landing. On one V.I. patrol which took me south to a point opposite the Thames estuary, I reported the launch of three V.2's from Holland, seen to be a red glow which would rise very slowly at first then accelerate rapidly to disappear out of sight. On sighting one of these I would turn immediately towards it and call my GCI station with "Big Ben" (the code for V.2's) giving bearing so-and-so. The GCI station would be able to plot a reasonable position line to find the launch site. On one occasion I reported a V.2 launch, then turned toward London where, a few minutes later, I saw the V.2 explode. This may have been an unique spectacle.

Church Fenton. We arrived here on 24 April 1945, and stayed until 20

November 1945. This period was in the dying days of the war and with little, or no enemy activity over Britain at this time, consequently the squadron entered a rather relaxed phase with crews on standby readiness and the occasional patrol. News from Intelligence was that a danger existed from a 'last ditch' stand from Norway-based Luftwaffe aircraft, and acting on this information two Mosquitos were detached to Acklington to cover the North of England against this threat. I was one of the pilots to take part in this action, but was scrambled only twice, on 2 May and 6 May, the last one being a 'bogey' which turned out to be a Wellington of 90 Group returning from a trip to Norway.

The war in Europe ended with an attendant great sense of anti-climax. Practice interceptions were carried out with some gunnery practice, and there were frequent cross-coutry flights to gain experience with our newly installed GEE radar equipment. Also, every crew made two long daylight trips over Germany and France to view the bomb damage and the airfields strewn with Hun aircraft. This was particularly interesting as all our operational flying had been done at night, when one sees very little. There was talk of postings to the Far East where the war was still raging, but I do not recall any of the squadron actually going. A considerable number of RAF ex-prisoners of war were repatriated to Church Fenton where they awaited the administration process. ++

Many of them wanted to get back into the air again and we made many flights taking them up in our Mosquitos. Much of our time was spent in squadron formation flying with displays given for various towns within reach, This culminated in the Battle of Britain display at the station, which was a great success, with 125 taking pride of place with their squadron drill and beat-ups across the airfield. During the immediate post-"VE" Day period there was considerable 'swanning' with the squadron Anson and Oxford, as well as Mosquitos being used to transport people all over the place - on private as well as official business. No really special events, nor visits by important personages can be recalled. There was however a well-known member of the then Labour Cabinet who turned up requesting to be taken to London most urgently. I stood by to fly her there, but the trip was cancelled.

I left the squadron at the end of October 1945, which just about coincided with the disbandment of the squadron".

Editors note: *Warrant Officer Radley was commissioned at Church Fenton in September 1945. He left the station to take up duties at No. 2 Ferry Pool, Aston Down, ferrying many different types of aircraft - singles, twins and fours. He was later selected for No. 8 course at the Empire Test Pilots School, Farnborough in 1949 and in 1953 became Deputy Chief Test Pilot with Hunting Aircraft (then British Aircraft Corporation), followed by a spell as Manager, Flight Safety and Accident Investigator for Saudi-Arabian Airlines. He logged over 220 types of aircraft flown in his long and full career in aviation, having started as an aircraft apprentice at Halton in 1938.*

In 1961, he became on of the first British members of the Society of Experimental Test Pilots, formed at Edwards Air Force Base, U.S.A. by a group of experimental test pilots including Brian Trubshawe, without doubt the most prestigious pilot society in the world and has very high membership requirements. All the original astronauts became members and just about all the subsequent ones. (that is, the test pilot astronauts).

++ See Air Vice-Marshal Plumtree's recollections in the Chapter "Station Commanders' Memoirs."

125 Squadron Church Fenton. 1945
W/O Barry Radley (right) with P/O Bert Taylor
and Mosquito NF Mk. 30
(Photo: Courtesy Barry Radley)

11

NO. 456 SQUADRON — ROYAL AUSTRALIAN AIR FORCE

(Press on - regardless)

The unit was formed at Valley, Anglesey, in June 1941, equipped with Boulton-Paul defiants and classified as operational on 5 September 1941. By the end of September, the Squadron commenced re-arming with Beaufighters — to be replaced in December 1942 by the Mosquito NF. II — the first patrol with these aircraft being flown 22 January 1943. The units future role was to be 'Night Rangers' and intrusion, interspersed with day sorties as long-range fighters . A move to Middle Wallop in March put the Squadron under the protection of No. 10 Group — fully operational. The following months were actively engaged in Ranger and Intruder sorties against enemy airfields and rail commuications — later to 'Instep Patrols'.

Via Colerne, Firwood Common and Ford, the squadron arrived at Church Fenton in December, 1944, under the command of Wing Commander Keith Hampshire, DSO (later DFC), who put the squadron through its most vigorous passage — although this was not realised at the time. In the words of Flight Lieutenant Ron Hodgen (recently here on holiday), "Wing Commander Hampshire was exceptional as a pilot, but a strict disciplinarian — the only trouble was that he expected us all to be of the same standard — which was very hard to attain — and maintain. He won the respect of all ranks".

Wing Commander Hampshire had had considerable experience of the war in the south west Pacific and in addition to other responsible posts, had commanded No. 22 Squadron RAAF which was equipped with Bostons. It was not long before the power of his mind and personality made itself apparent. He inaugurated a regime of thoroughness which, coming as it did after a long period of virtual inactivity, seemed burdensome at first, but before long gathered momentum and carried the Squadron along in a wave of sustained and concentrated activity which placed it among the most successful fighter squadrons in the service. He conceived his purpose in the simplest terms; to weld the Squadron into an efficient fighting unit to enable it to discharge the responsible task that was to be assigned to it in the spring and summer of 1944. He carried out his purpose by the power of his example and the vindication in practice of his methods of infinite care right down to the most meticulous detail. He made the Squadron realise its organic unity — the interdependence of air and ground crews — and drew every member into the combined effort by giving him a sense of his function and responsibility, by providing opportunities for pooling ideas and, above all, by infusing a new sense of drill and discipline. It is neither excessive nor invidious to say that he contributed more to the making of the Squadron without a certain amount of stress and friction. He was working against time and had to overcome the "vis inertiae" which was the heritage of months of easy going comparative idleness. But he did communicate something of his dynamic energy and his technical mastery of the difficult art of combat flying. If he made some mistakes, if at times he seemed to be the maker and breaker of his own laws, he was intensely interested in every man in the unit, and was scrupulously just. He never spared himself or anyone else, where the efficiency of the Squadron was concerned. When at length they found themselves in action, the men realised the value of his leadership and no member of the Squadron could have had the slightest doubt that he was serving under a great commander. Even omitting his own performances, which won him the DFC and a Bar to his DSO, the record of the Squadron in the ensuing months would prove to be the measure of his achievements.

An increased state of readiness was maintained on Christmas Eve as a precaution against the enemy's disposition to mount attacks on occasions of national festivity and relaxation. However, there was no activity.

It is appropriate at this point to mention some more general matters outside the strict limits of the internal history of the Squadron; matters which were none the less to give a decided twist to that history and to open up new spheres of activity. This judgment must not be taken absolutely. It is true that the year 1944 provided opportunities for night fighting such as never had been available before, but the results achieved were not merely due to the opportunities. The two and a half years of apprenticeship had been of incalculable value.

It was becoming obvious that the long period of small spasmodic raiding by the enemy, which had lasted since the decline of the first concentrated night offensive in the middle of 1941, was giving place to more vigorous and novel methods. This foreshadowed the possibility of a renewed night offensive in strength. Throughout the autumn and early winter of 1943/44 the fighter-bomber night attacks became stronger and more frequent. The long-range bomber forces in the west showed signs of increase. Moreover, two new technical devices were discerned as being applied by the enemy; one a form of pathfinder technique and the other the use of bundles of metal foil or metalised paper strips dropped by raiding aircraft for the purpose of jamming both ground and airborne radar. This device was codenamed 'window' by the British who became the biggest users of it.

For the opposing forces of the RAF this period was also a time of change.

Fighter Command disappeared to be replaced by a new organisation entitled Air Defence of Great Britain (ADGB) and in the night fighter squadrons, re-equipment was soon to take place. As far as No. 456 Squadron was concerned, this was a long overdue move. New aircraft were required as was a more adequate airborne radar. The aircraft strength had all flown several hundred hours and were coming due for major inspections and engine changes. As to the radar, the AI Mk. IV and V sets were becoming obsolescent and the Squadron had been passed over when the improved AI Mk. VIII was introduced. However, an even better type of radar, the SCR 720 or AI Mk. X, of American design, had been evolved and the Squadron had been listed high among the units which were to have it when ready. This was to prove very fortunate, for its performance not only greatly surpassed that of any of its predecessors but it proved to be the type least embarrassed by interference from Window. It provided a much wider scan, it gave good results at low level and could be equally well employed for controlled or freelance interceptions. The latter became very important when ground stations became saturated by Window.

Mention has already been made of the work done in, and for, the Squadron by Group Captain Hampshire. At the time of his departure he gave a party to all ranks and ended his speech with that exhortation which had ever been his own gospel and had become indelibly impressed on all who had served with him — "Press on — regardless".

The concluding phase of the Squadron's life-history began with the move to Church Fenton. At the time it seemed to indicate the opening up of new fields of action and exciting possibilities. Hitherto, night fighters of the RAF had essentially a home-based role, extended by means of Intruder and Ranger sorties over the Continent. The liberation of France had enabled night fighter squadrons of the Allied Expeditionary Air Force to be based on Continental aerodromes and to operate in immediate support of the army battle fronts. During the transition stage the Squadron had been able to share in the Continental patrols over the Low Countries by operating from advanced bases such as Manston. Now, however, it was diverted to a completely different task, for which a certain amount of preparation was necessary. By this time the Squadron had become operationally trained for this new role, the tide of war on the Continent had swept so far and so rapidly that in the end very few opportunities were to be had. The Luftwaffe was palpably failing and the battle fronts were getting almost beyond the range of home-based fighters. Hence the operational record for the period is comparatively meagre. Although good work was done, there are but a few spectacular incidents to enliven the narrative.

The remnants of the enemy long-range bomber forces were no longer a menace to the United Kingdom. They were sufficiently occupied with the land-battle fronts. Precautions had to be taken still against the Heinkels launching flying Bombs from the North Sea, but their attacks were spasmodic and on a small scale. Far more effective were the long-range rocket projectiles, and against them no interception methods could be applied. There was one serious threat — tactical rather than strategical — that had to be guarded against by home-based night fighters, namely efforts by the Luftwaffe to intrude into the air space of Bomber Command bases in eastern England. It had been known for some time that the Germans had been making elaborate and careful plans for the use of their own night fighters — their one remaining force of numerical strength and high efficiency — for this type of operation, and it was against this force that the available RAF night fighters were to be deployed.

The new work for which the Squadron was destined was long-range fighter escort on support of Bomber Command operations under the control of No. 100 Group. This Group specialised in a great variety of counter measures designed for direct and indirect assistance to Bomber Command, and had been functioning since November 1943.

At Church Fenton the Squadron completed re-equipment with the Mosquito NF Mk. 30. It was a more powerful aircraft, specially designed for best performance at higher altitudes. It was fitted with AI Mk. X, Gee, Monica, long-range fuel tanks (drop tanks could be fitted), and four 20mm cannon. Flying training consisted mainly of cross country flights to exercise navigators in the use of Gee. As most of the operations would be in areas far beyond the range of ground control facilities, it was necessary for crews to be able to look after themselves and to find their objectives and patrol lines independently.

After leaving Ford the first experience of the Squadron at Church Fenton was considerable physical discomfort. The Station itself was pleasant enough, with good buildings and adequate facilities for maintenance. However, at the time it was over-crowded and quarters and dispersals were anything but satisfactory. Moreover it was mid-winter.

Six aircraft were required at readiness nightly. The state was raised to eight for a few days, but then went back to six when two aircraft and crews were detached for duty at Ford. The general operational commitment of the Squadron in the north Midlands area was to counter enemy aircraft and flying Bombs; the two aircraft at Ford were to operate in conjunction with other units on anti-Heinkel patrols in the south east area. The first patrols were flown on 3 January 1945 and continued throughout the month without incident. The sorties were controlled by Patrington GCI, Dimlington CHEL (Chain Home Extra Low) Stations and Digby Sector Control. There were few scrambles, but on no occasions were any contacts or sightings obtained. A total of 88.50 hours was flown during the month, of which 45 were operational.

Also during January two experienced crews, Flying Officer Sanderson/Pilot Officer Nicholson RAF and Flying Officer Hodgen/Warrant Officer McCormick RAF were detached to Fighter Interception Development Unit at Ford for special duties in connection with an airborne equivalent of GCI.

Changeable weather restricted flying and caused sickness among aircrews, making it difficult to find sufficient numbers to maintain the required state.

Bad weather continued into the next month, but none the less, practice cross country flights were organised and anti-intruder patrols were flown on the first seven nights. There was no enemy activity. On the first two nights a state of ten aircraft at readiness was required. On the third it was reduced to eight. By the 9 February the Squadron was released from operational flying for training purposes. As many aircraft as could be made available were to be devoted to the night cross country practice flights until the navigators could be fully trained. For the rest of the month this was the sole activity when weather permitted. The only excitement were some troubles with the new engines of the Mark 30 Mosquitoes. On two occasions single-engined landings had to be made.

Living conditions improved during the month, mainly due to the generous policy of the Station Commander, Group Captain Hawtry, who like the other Station Commanders already mentioned, won the gratitude and respect of the Squadron. At the end of February official notification was received that the unit was to move to Bradwell Bay on 3 March.

However, the last night at Church Fenton was to provide a scene unparalleled in the annals of the Squadron. The main party had left for Bradwell Bay by train, leaving behind the aircraft and crews to follow on next morning, after the customary celebrations. At the very height of festivities (at approximately 1 am on the morning of 4 March) the enemy sent over between 50 - 80 aircraft on intruder activities in the neighbourhood. The attack achieved surprise and some measure of success against returning bombers. The Squadron had been stood down and had no operational commitment at the time. Group HQ rang to ask if anything could be done. Although there were no starter batteries or ground crews, within an hour five aircraft were scrambled. There was a general exodus from the Mess as the crews raced down to the Flights. With the assistance of whomever was available — the Squadron Doctor was seen acting as fitter to the Wing Commander's aircraft — the five crews became airborne. They were: Wing Commander Howard, /Flight Lieutenant Ross; Squadron Leader Cowper/Flight Lieutenant Watson; Squadron Leader Samson/Flying Officer Abbey; Flight Lieutenant C. C. McLennan/Flying Officer Fraser; Flying Officer W. A. McLardy/Flying Officer R. A. Woodman. The aircraft had only one channel on the radio transmitter, nevertheless they patrolled for three hours. Flying Officer McLardy obtained a visual on a Ju 188, fired a short burst but missed. The Wing Commander also had a chase, but was unable to close in. The whole thing was considered a creditable performance under the circumstances, and the Squadron received a congratulatory message from the Senior Air Staff Officer of No. 12 Group. It is important to note this, for otherwise the events of this night should perhaps be passed over as unworthy of the dignity of the history. Actually it was quite a serious business, though the Squadron emerged with credit. It had a humorous side too, as witness the case of a certain observer, who seeing the transport about to leave the Mess, clambered in thinking it was merely a detachment going to the local hostelry. Before he could realise what was going on he found himself airborne.

When day dawned there was chaos of difficulty with no comic relief. In the morning instructions from Group were that the aircraft were not to leave until further notice. Shortly after lunch the move was cancelled completely, and the train recalled from Bradwell Bay. A state of ten aircraft was required. Fifty ground crew with essential equipment were flown back by Dakotas, and a petrol refueller was borrowed from Bomber Command.

That night - 4 March - enemy activity was renewed on a smaller scale. Patrols were flown by five of the Squadron aircraft. Two made interceptions on the enemy. Flying Officer Hodgen obtained a visual on a Ju 188. It was recognised from 400 feet at a height of 1,500 feet. The pilot aimed at the port engine and fired a short burst. Strikes were seen on the engine and wing, but the vibration of the cannon firing caused the gun-sight to fall to pieces exposing the lamp and dazzling the pilot, who lost the visual. The navigator saw the enemy aircraft go down to starboard, but a further search proved unavailing. It was claimed as damaged.

Flight Lieutenant H. Smith also intercepted a Ju 188 and identified it at about 100 feet range. It was very dark at the time and the Mosquito overshot on the approach. When contact was regained the enemy dived down to 150 feet above the sea. The chase lasted for 40 minutes, but the visual was not regained.

The Squadron was beginning to settle down to normal routine again when the order to move was revived. This time it was carried through and the Unit commenced operating from Bradwell Bay on 17 March, where it was disbanded on 15 June 1945.

Wing Commander Hampshire moved on to Transport Command.

(Editors note: This narrative was specialy prepared by the RAAF Historical Section, Canberra)

'A' Flight — No. 456 (Australian) Squadron (with Mosquito)
Ron Hodgen — second from left — front row.
(Photo: Courtesy Ron Hodgen via RAAF)

Wing Commander Hampshire with his navigator Flight Lieutenent Condon.
(Photo: Courtesy via T. Condon/D. Vincent)

12

No. 609 (WEST RIDING) SQUADRON

No. 609 Squadron was formed at Yeadon, Yorkshire on the 2 February, 1936 as a Day Bomber Squadron unit of the Auxiliary Air Force in No. 1 Air Defence Group. The Commanding Officer was Squadron Leader Harald Peake, the aircraft on strength were Avro Tutor and Hawker Harts — later the Hawker Hind. The site of the airfield is now occupied by the Leeds & Bradford Airport and this is commemorated by a gleaming wall plaque unveiled in March 1968. To those present, Leeds & Bradford Airport will always be Yeadon.

On the 8 December 1938, the Squadron was transferred to Fighter Command and Squadron Leader G. H. Ambler assumed command. The later careers of both these Commanding Officers was rapid. Squadron Leader Peake became Director of the Auxiliary Air Force with the rank of Air Commodore, whilst Squadron Leader Ambler rose to become the only Auxiliary Officer in the war to attain the rank of Air Vice-Marshall when appointed Senior Air Staff Officer at H. Q. Fighter Command.

In July 1939, 609 Squadron spent two weeks at Church Fenton for their annual training. With the outbreak of hostilities on 3 September 1939, 609 was stationed at Catterick, North Yorkshire, equipped with Mk. I Spitfires. Its first engagement with the enemy occurred on the 29 January 1940, when a section of the Squadron then, operating from Drem, chased an He. III which had been attacking shipping. Its first success however, did not come until the 27 February when a German bomber of the same type was destroyed when raiding a convoy.

After a period of night flying, the Squadron was moved to Northolt on 21 May and played a prominent part in the air fighting over the Dunkirk area during the evacuation. Following the evacuation from France, No. 609 Squadron was engaged in the defence of London during the Battle of Britain. In the course of intensive fighting in August, no less than 46 German aircaft were destroyed by the Unit. By 21 October the hundredth enemy aircraft fell to the Squadron when Flight Lieutenant Howell chased a Ju. 88 at tree-top level until it hit the ground. 609 Squadron was the first Spitfire Squadron to shoot down 100 enemy aircraft. Flight Lieutenant John Dundas, DFC and Bar, shot down Helmuth Wick on the 28 November 1940. He was one of the highest scoring Luftwaffe officers at this time. John Dundas was almost immediately shot down and killed by Wick's No. 2.

The Squadron moved to Warmwell on 29 November 1940.

A quiet period followed, with a few scrambles and interceptions, until 25 February 1941, when a move was made to Biggin Hill. There the Squadron formed part of a Fighter Wing which carried out sweeps over Northern France, and also provided convoy escort and standing patrols.

The Squadron continued to add more enemy planes to the number already shot down, numerous interceptions and combats taking place. A particularly good day occurred on 8 May, 1941, during a patrol over the Channel — see combat report entitled 'Battle of the Dinghy'.

In June the role of the Squadron changed from the defensive to the purely offensive, and large scale sweeps were made.

The unit fought with great distinction, operating from a variety of bases, taking part in sweeps over France against targets, usually marshalling yards, power stations and constructional works.

March 1942, was the last time Spitfires were flown on operations — the Squadron was re-armed with Typhoons in May. Intruder patrols into enemy territory commenced in November 1942 against rail communications, the chief scorer being the Commanding Officer at the time, Squadron Leader R. P. Beamont, who destroyed 20 locomotives, earning a well deserved bar to his DFC.

By the end of March 1944, the squadron tally of enemy aircraft destroyed was 231, and they became part of the 2nd Tactical Air Force. Moving to Wunsdorf on the 27 May 1945, the unit was reduced to a number only, and all personnel posted away, to be reformed at Church Fenton on 10 May 1946, equipped with Mosquito NF. 30 night fighters. In the Spring of 1947 the squadron moved to Yeadon — its first home. In April 1948, the squadron converted to Spitfire XVI's.

Among the decorations awarded to the squadron were: 3 DSO's . . .22 DFC's . . .and 4 DFM's. Also several Bars to the DFC.

In January 1950, for the first time it its history, the squadron was awarded the Esher Trophy, awarded annually to the Auxiliary Squadron judged to be the most efficient. In 1954, the squadron were the runners-up to No. 615 Squadron.
On March 1957, this squadron, along with other Royal Auxiliary Air Force Squadrons, was disbanded.

INTELLIGENCE COMBAT REPORT

'BATTLE OF THE DINGHY'

Sector Serial Number	(A)	IG/211
Date..	(C)	8/5/41
Flight, Squadron	(D)	Flight: A & B Sqdn: 609
Number of Enemy Aircraft...........................	(E)	Approx. 15
Type of Enemy Aircraft..............................	(F)	Me. 109
Time attack was Delivered...........................	(G)	1715 — 1730 hrs.
Place Attack was Delivered...........................	(H)	Strait of Dover.
Height of Enemy	(J)	Mostly below 6000 ft.
Enemy Casualties..	(K)	6 Me. 109 destroyed 2 Me. 109 probable.
Our Casualties, Aircraft..............................	(L)	1 Spitfire Cat. 2.
Personnel...	(M)	Nil
Total number of rounds fired:		9,286

General report

12 aircraft of 609 Squadron (Spitfire 2 "303" Browning) took off from Biggin Hill at 16:49, 8/5/41 with orders to patrol Maidstone at 15,000 ft. They were then told of a dinghy containing a German aviator in the sea about half way across the channel from Dungeness. 609, who were in three sections of four, were given a vector of 100º and were losing height when they spied a rescue boat on fire with 2 Me. 109's circling it. Leaving Blue section as top cover, 609 dived out of the sun, but E/A disappeared. Blue section was attacked and Blue 3 (Sergeant Mercer) was hit, but, turning sharply fired at one of the two E/A which passed him from his port side. He delivered beam, head-on and stern attacks. The last E/A emitted smoke and was seen diving vertically for 5000 ft. before it was lost sight of. It is claimed as probably destroyed. Oil from its engine was found later on the wing of Blue 3's machine.

Red and Yellow sections, reforming at 10,000 ft. over Dungeness, were told of a second rescue boat proceeding east. Me. 109's were seen circling this at 2000 ft. and others were in the vicinity. Leaving Yellow section as top cover, Red section dived to attack and were joined by Blue 1 and Yellow 4. Red 1 (S/L Robinson DFC) attacked the nearest Me.109 and shot it into the sea after a 5-second burst; 800 closing to 200 yds. Red 2 (F/Lt. Curchin, DFC) fired two bursts at close range at another which emitted smoke, turned over and was finished off by Yellow 4 (Sgt. Hughes-Rees). Red 2 then attacked another E/A and after a long burst from astern at close range, pieces flew off E/A, its port wing buckled and it dived into the sea. Red 2 was then hit by cannon shell and made for base.

Two more Me. 109's were shot down in quick succession by Blue 1 (Sgt. Rigler) from point blank range. The first was seen diving vertically 200 ft. from the sea when he engaged the second which crashed in flames, and immediately afterwards a patch of green oil was seen in the expected spot near the burning wreckage of the second, both of these objects being seen by Yellow section. Blue 2 (Sgt. Palmer) pursued 2 E/A within 5 miles of Calais, but was unable to get within effective range, although he fired three 1-second bursts.

Meanwhile, Red 1 (S/L Robinson) had pursued his second Me. 109 and caught it up 5 miles from the French coast, shooting it down into the surf offshore, the A/C sinking. At the same moment he was attacked by about nine Me. 109's from all directions. He called for help and flew back towards England at top speed and at 0 feet, taking superb evasive action by turning towards each E/A as it opened fire. His machine was undamaged. Blue 2, hearing his call for help, saw a Spitfire pursued by a Me 109 and attacked the latter with two 2-second bursts from 100 yds. and then point blank range. E/A's propeller appeared to be stopping and it sped towards the sea. Blue 2 was unable to follow it as he himself was then attacked. It is claimed as probably destroyed.

Yellow section report that while flying at 8,000 ft. they were attacked by a pair of Me. 109's from 15,000 ft. R/T of Nos. 1 and 3 had become U/S and No. 2 (P/O Malengreau, Belgian) turned above in front of leader as a visual warning and fired short bursts without effect at 1 E/A as it flashed past, while Yellow 1 and 3 made quick evasive turns. Yellow 1 reports he saw a large speed boat proceeding rapidly towards France from the burning vessel.

All our aircraft landed safely at Biggin Hill by 17:55 hrs.

Editor: The format and service abbreviations are as authentic Report.

Sir Trafford Leigh-Mallory and Air Commodore Harald Peake, with members of 609 Squadron
(Photo: Courtesy of Mr. Chris Goss)

609 Members - and Mosquito's (Photo: Courtesy Yorkshire Post)

An altogether — Better 'Approach'
FAST & FRIENDLY
NORTH
LEEDS/BRADFORD
SOUTH
+ M62 M63
WELCOME TO L·B·A
FOREIGN
DEPARTURE LOUNGE DOUBLED IN SIZE
PASSENGERS UP
DUTY FREE UP
TRAFFIC UP
PROFITS UP
FREIGHT UP
HEADACHES DOWN
Leeds Bradford

IMPRESSIONS OF "609"
by
Flight Lieutenant S. H. (Darkie) Hanson - MBE

It is true to say that the Auxiliary Air Force formed part of the 'cutting edge' of the Royal Air Force in World War II, since they represented 21 fighter squadrons of Fighter Command at the outbreak of hostilities in 1939. For this reason it is of appropriate historic interest to describe the build-up, and necessity, for these squadrons.

The termination of World War I, with the Armistice in 1918, saw general de-mobilisation of all combatants. The Royal Air Force was subjected to a general run-down to become a 'shoe-string' service, but still high in spirit, pride and resolution'.

Under the aegis of World War I veteran, Air Marshall Lord 'Boom' Trenchard, subsequent 'Father of the Royal Air Force', the fortunes of the RAF were in capable and dedicated hands.

Despite the Governments parsimonious attitude towards the rebuilding and modernising of the country's youngest service, in April 1922, the new RAF Staff College was established at Andover — to be followed shortly by recommendation to create 15 new squadrons for a Home Defence Force — proposed by the Committee of Imperial Defence. Motivated by the increase in France's military aspirations, at this time numerically the strongest in Europe, in June 1923 the new Prime Minister Stanley Baldwin announced the Government's proposals to create a Home Defence Force of 52 new squadrons — an increase of 34 units to the existing force, to be operational by 1930.

Beneficiaries of this new programme would be the Royal Air Force and Fleet Air Arm, which in 1925 became a unified Command, having the title of Air Defence of Great Britain (A.D.G.B.) This new command was nurtured and developed under the very experienced Air Marshall, Sir John Salmond.

Of the proposed 52 squadrons, 35 would be bomber and 17 fighter units — six of which would be Auxiliary Air Force squadrons, manned by part-time citizens, similarly organised to the existing Army Territorial Force. Inaugurated in 1925, its first squadron was No. 602. From 1926, the A.D.G.B. operated in two basic sections (a) The Wessex Bombing Area (Bomber Units) and (b) The Fighting Area (Fighter squadrons). In 1927 the Auxiliary Air Force (A.A.F.) and its special reserve, were designated No. 1 Air Defence Group and incorporated into A.D.G.B. Due to the policy of continued expansion, the A.D.G.B. concept was scrapped in July 1936, to be replaced by four new and interdependent commands - Fighter, Bomber. Coastal and Training Commands.

As part of the new Fighter Command, No. 609 A.A.F. Squadron was established in 1936 and based at Yeadon — geographically conveniently located between Leeds and Bradford — on the site now occupied by the commercial airport.

The Honorary Commanding Officer was Harald Peake. The squadron operated mainly at week-ends with a programme of flying training using Hawker Hart and Avro Tutor aircraft, and ground trades (engines and air frames). Included in the programme was a two week camp in August.

With great enthusiasm and a professional background in mechanical engineering, I arrived at Yeadon with aspirations to partake the course leading to Sergeant-Pilot. However, my hopes were temporarily dashed when the Commanding Officer said a firm "No" — qualifying this by stating that I would not be considered for air crew training until I had successfully passed my Ground Trade examinations — with the appropriate rank of L.A.C. (Leading Aircraftsman).

Since my professional background and leisure pursuits dictated, I chose to take the course most dear to my heart — Fitter Aero Engines (F.A.E.). My aspirations for Sergeant-Pilot did not materialise, for reasons explained later.

By 1939 the Squadrons of the Auxiliary Air Force were embodied in Fighter Command, later to serve with distinction in the war. In recognition of this, the late King George VI conferred the prefix "Royal" on the Force. The squadrons were affiliated with the cities or counties in which they were located — from which they took their title, eg. No. 609 (West Riding of Yorkshire) and No. 616 (South Yorkshire) Squadrons. Each unit was responsible for its own administration, instruction, recruiting, etc, and a Regular RAF Adjutant was appointed, with a Regular cadre to assist.

When war was declared in September, 1939, I was already a trained Leading Aircraftsman (F.A.E.) and involved in my ground trade on aero engines. The squadron was mobilised, at this time being stationed at Catterick (North Yorkshire).

I was now permanently 'grounded' and in 1943 had attained the rank of Sergeant Fitter (2E) and regarded in the squadron as a 'Typhoon Specialist' — our current tank-busting aircraft. By this time my experience as a Technical Officer was considered too valuable to warrant any further thought of transfer to air crew training. Consequently, my original pipe-dream as Sergeant-Pilot was now a non-starter. I was however, very happy and satisfied with the situation and highly regarded in the squadron.

The squadron had been heavily involved in the air-war since the onset of hostilities, in the north and south of the country. We were highly mobile, as were most of the squadrons of Fighter Command.

During the period July 1944 to May 1945, the squadron was operational in Europe as part of the Allied 2nd Tactical Air Force, during which time the squadron lost several Commanding Officers. One of these was the very popular, charming Squadron Leader Pat Thornton-Brown. A short time before Christmas 1943, he and five other pilots flying Typhoon Mk I, were appointed to act as close support escort to three formations of American Marauder bombers — two Typhoons to each formation. They were attacked without warning from astern by numerous American Thunderbolt fighters. Despite recognition signals, given in vain, two Typhoons were shot down — Chuck Miller, a Canadian, and the Commanding Officer Pat Thornton-Brown. Another Squadron Commander, Squadron Leader T. Y. Wallace, a South African, was shot down over Dunkirk in November 1944 — the Typhoon crashing on the sands of Dunkirk. Years later his mother visited the scene of the crash and remarkably, following a storm in the region, the tail-fin of a Typhoon was visible. On digging it was established that it was the aircraft belonging to Squadron Leader Wallace, with the pilot seated in the cockpit!

In 1945 the defeat of the German forces was complete, celebrated by V-E Day. 609 Squadron was disbanded at Wunsdorf and reformed at Church Fenton, operating from Yeadon as a Night fighter squadron. They eventually returned to Church Fenton in October 1950 as a Day Fighter squadron, equipped with Spitfire 16 aircraft, to be succeeded by the Vampire FB.5 and Meteor F.8.

At this stage the Sector Commander was Air Commodore John Grandy (later Marshall of The Royal Air Force — Sir John Grandy).

In 1951, I was commissioned, eventually attaining the rank of Flight Lieutenant, and in 1957 was honoured to receive the MBE in the New Year Honours list. It was a proud occasion when the squadron won the Esher Trophy in March, 1955. This award is presented to the most efficient squadron in the Auxiliary Air Force.

On 4 March 1957, the '609 Air Force' was disbanded and the full squadron paraded at Church Fenton for the last time — with the huntsman of the Bramham Moor Hunt sounding the squadron's motto — "Tally Ho" — an emotional and historic moment for all present. For me, this was the close of a hard, but privileged and glorious association with a great squadron, redolent with memories of actions, aircraft and comrades-in-arms (ground crew and aircrew), some having paid the supreme sacrifice. Given the opportunity, I would very willingly repeat the experience.

S. H. Hanson December 1988

Group Captain Douglas Bader in Spitfire DB Tangmere (Photo: Courtesy G/C Ken Holden)

Group Captain Ken Holden with his personal Spitfire.
(Photo: Courtesy G/C Holden)

13

No. 616 (SOUTH YORKSHIRE) SQUADRON

of the
AUXILIARY AIR FORCE

(Motto: No rose without a thorn)

Frequent visitors to Church Fenton to take part in ground and aerial exercises were aircraft of No. 616 (South Yorkshire) Squadron. They also took part in the fierce, bitter battles over the now historic beaches of Dunkirk and were also among those who patrolled the Normandy beach-head four years afterwards when British troops were once more fighting in Western Europe. Later it became the first RAF squadron to operate jet propelled aircraft against the Luftwaffe,

The squadron served with distinction throughout the Battle of Britain and during the period when an offensive air policy superseded that of defence.

No. 616 Squadron was formed at Doncaster on 1 November, 1938, with Squadron Leader, The Earl of Lincoln as Commanding Officer. That of course, was the time when the Munich crisis had barely passed into history and men from the Doncaster, Sheffield and Barnsley districts lost no time in volunteering for a service which was the RAF's counterpart of the Territorial Army.

It is interesting to note that the first twenty pilots were mainly engineers with managerial jobs in civilian life; men from Doncaster's railway works were prominent among those who joined the new squadron as its ground mechanics.

When war was declared the squadron was at its first annual camp at Manston, Kent, and those who had been "week-end" airmen found themselves among the nucleus of a force that must do battle with Göering's much-vaunted Luftwaffe. A period of hectic training followed, but in May 1940 the squadron was ready for action and moved to Southend to relieve No. 74 Squadron.

By then the Germans were sweeping on into France and the Low Countries, and the end of May saw the heroic last-ditch stand and the evacuation of the British Expeditionary Force from Dunkirk. Along with No.'s 65 and 19 Squadrons, 616 was given the job of patrolling the beaches, flying from dawn to dusk to protect the ships evacuating our troops and engage in combat enemy forces of overwhelming size.

First blood was drawn when Flight-Officer Moberly — lost five months later in a battle with a hundred enemy fighters — shot down a Me. 109. During the fights over Dunkirk, 616 destroyed five and damaged many more German aircraft.

The Battle of Britain followed soon after the disasters on the Continent. It was a new kind of battle — the first of its kind — in which two air forces thrown at each other independent of land forces, fought for the ultimate possession of an island country. In the end it was won by men who only twelve months before had been engineers, merchants, students and clerks who now fought alongside the more experienced men of the regular RAF.

There was another difference. Unlike Waterloo and Trafalgar and other momentous battles of the past, the Battle of Britain was visible to thousands of townsfolk and countrymen alike as they went about their daily tasks. It was a battle in which the people of a country, no less than the men who fought it, were in the front line day after day.

Unlike other auxiliary squadrons who fought the whole of the Battle of Britain away from the districts from which their personnel originated, 616 did at least fight the opening period off Yorkshire's coast. The Luftwaffe sent a formation of Ju. 88s to bomb a Yorkshire bomber airfield. 616 intercepted them off Flamborough Head and dived into the heart of its formation. As the result of that afternoon's work, eight Ju. 88s were shot down and six others probably destroyed and damaged.

On February 1 1941, the South Yorkshire squadron moved to Tangmere where its leader was the man whose name will ever be associated with the story of the Spitfire, and the late British fighter-pilot, Group Captain Douglas Bader DSO and Bar, DFC, known the world over as "the legless pilot". He lost both his legs in a flying accident at Woodley, Reading in 1933.

Within nine months of the accident he was equipped with artificial legs, took a course in civil aviation and piloted a plane, operating the rudder controls with his artificial limbs. His application to be taken back into the RAF however, was refused. On the outbreak of war he renewed his application to the Air Ministry. This time he was accepted as a fighter pilot.

Among those who earned special distinction was a young Cawthorne (near Barnsley) pilot, H. S. L. Dundas, who was later awarded the DFC. The citation accompanying the award said: "This officer has shown unflagging courage in the face of the enemy and the utmost tenacity in supporting his leader. He has destroyed at least three enemy aircraft and damaged others".

In July 1944, the squadron moved to Manston, where its war career had commenced. Already the first flying bombs were penetrating the lushness of the Southern English countryside, bringing death to villages, townships and cities. Once more the people of Kent were 'under fire'. Flying Spitfires first and later Meteors, 616 Squadron again went into action against a menace from the air. By August it had destroyed $13^{1}/_{2}$ V-I's. On August 26, 1941, the late Air Chief Marshal, Sir Trafford

Leigh Mallory presented the squadron with its crest in the form of the White Rose of Yorkshire pierced by an arrow denoting speed and death. Underneath was the motto: "Nulla Rosa Sine Spina" (No Rose Without a Thorn).

From the beginning of hostilities until the German surrender that thorn never missed an opportunity of piercing the enemy. The squadron was disbanded in March 1957.

Dispersal at Tangmere — with Nip Hepple (2nd left) Douglas Bader (with pipe) and Ken Holden (standing at right) (Photo: Courtesy G/C Holden)

"Time for ralaxation" — Ken Holden, Buck Casson and Hugh Dundas (Photo: Courtesy G/C Holden)

KEN & EUSTACE HOLDEN

(a biography of two outstanding brothers)

Some 30 miles South of Church Fenton lies Doncaster and nearby Royal Air Force station Finningley — historic home of No. 616 (South Yorkshire) Squadron of the now disbanded Royal Auxiliary Air Force — the 'week-end' Squadrons.

A product of this Yorkshire town were two talented brothers — Ken and Eustace Holden. Ken studied motor engineering at Leeds College of Technology and started in business as Ken Holden (Ltd.) Automobile Engineers. Meanwhile, Eustace joined the RAF as a 'Regular ' and was commissioned as a fighter pilot.

Group Captain Ken Holden DFC, OBE. (Photo: Courtesy G/C Holden)

Influenced by brother Eustace's successful career in the RAF, and the impending threat of war, Ken decided to enlist in Doncaster and shortly afterward found himself with No. 616 Squadron at Finningley in February 1939. The Squadron Leader was the Earl of Lincoln, under whose aegis the unit was rapidly developing. Early flying training was carried out with Tiger Moths, Avro Tutor and Avro 504N basic trainers. Ken flew 'solo' in a Tutor on 26 March, and was commissioned as an acting Pilot Officer in April. Further flying was programmed using more advanced aircraft in the Gloster Gauntlet and Hawker Fury — later the Gladiator. Summer camp was spent at Manston, Kent, and it was about this time that the AAF was embodied into the Royal Air Force.

The war clouds gathered and on 1 September Hitler invaded Poland without warning or provocation. Honouring our Treaty with Poland, Britain declared war on Germany at 11.00 am on 3 September. 1939, followed one hour later by France.

General mobilisation followed immediately and 616 moved to Leconfield, a new war base near Hull, and a satellite station of Church Fenton. Ken collected the first squadron Spitfire Mk.I (K9-804) from Duxford. On 24 November came a successful assessment as a fighter pilot and the posting to II Group Fighter Pool, St. Athen, later to No.2 Ferry Pilot Pool at Filton, Bristol. Returning to Leconfield on 2 January 1940, the squadron took part in extensive flying exercises with squadrons of the sector base of Church Fenton, followed by convoy patrols over the East coast. April saw the fitting of a much-needed safety feature to the Spitfires when armour-plating was adapted to the underside and rear of the pilot's seat.

On 28 May squadron aircraft covered the evacuation of the depleted British Expeditionary Force at Dunkirk, when Ken shot down his first enemy aircraft, a Bf 109 — plus another unconfirmed. 616 lost three aircraft from a force of twelve; Mae West's and service revolvers were carried for the first time.

On an offensive patrol on 1 June over Dunkirk, Ken shot down two 109's plus one unconfirmed, but due to damage sustained during the encounter to his Spitfire he had to make a 'dead prop' forced landing at Colchester barracks. During a enemy raid on Hawkinge, he damaged two further 109's.

On a short stay at Coltishall he met the legless Douglas Bader, when new tactics were discussed. This was the start of a long and enduring friendship.

During this period the squadron was extremely transient, with bases as widely dispersed as Kenley, Coltishall, Kirton-in-Lindsey, Tangmere and Westhampnett. At Kenley, the squadron lost seven pilots taking off to intercept enemy raiders — 36 un-exploded bombs lay around the airfield after the attack. From Kirton-in-Lindsey during September, patrols were made over London and the south-east coast.

By this time replacement pilots had joined the squadron, which included Pilot Officer's Johnnie Johnson and 'Nip' Hepple. Ken was promoted to Flight Commander with the rank of Flight Lieutenant. Personal tributes to Ken from both of these (then) inexperienced young pilots.

Pilot Officer Johnson (now Air Vice-Marshal) remembers - *"I was introduced to some of the original Auxiliary members of the squadron and was immediately impressed by a thick set officer with a ruddy, expensive complexion, who seemed considerably older than the rest of us. This character was Ken Holden, who had played rugby for Yorkshire on many occasions - and looked the part. Ken had learned to fly at the ripe old age of 29 and during the following years I came to know him well. Whatever the state of the war, he always insisted on his Saturday evening excursions, which consisted of changing into a good uniform immediately after flying was over. We then drove to the nearest sizeable town with plenty of good ale, followed by a variety show, dinner - and more ale! Ken was a good mimic, always full of humour and a most lively and entertaining companion. But during our first weeks together, this side of his character was not revealed and I thought he was inclined to be dour and cautious. His north-country wisdom tended to frustrate us younger chaps, who soon thought we knew all the answers and were dead keen to get at the Huns. Later, Ken became our Flight Commander and some of us who saw the war through owe our necks to his thoroughness and the lessons we learned from him".*

A similar tribute to Ken was made by Pilot Officer 'Nip' Hepple (later Squadron Leader - who commanded No. 229 Squadron in Malta and No. 118 in the U.K.)
"Ken Holden, our direct Yorkshire Flight Commander, with his 'now you young lads' fatherly approach, probably did much for my career with his careful guidance and advice".

In March 1941, the Tangmere wing was formed under Douglas Bader (ex 242 Squadron) comprising Spitfire squadrons 145 - 610 (City of Chester AAF) and 616, 610 using the satellite airfield of Westhampnett. Douglas Bader was very keen on aerobatics and shortly after his arival was soon in the air giving almost daily displays in his Spitfire, to the delight of the ground crews - and the squadron in general. He had complete confidence in his aircraft and contended that no pilot was the master of his aircraft until he could control it in all attitudes. Sweeps over France were made, with fierce encounters with the enemy. 'Rhubarb' operations (low level sorties usually in bad visibility and low cloud) were frequently carried out in the search for opportunist targets - staff cars, locomotives, enemy troops, rolling stock, stationary aircraft etc. However, it was a widely held opinion by some pilots that the benefits of this type of operation fell short of the losses to pilots and aircraft. Soon Ken was promoted to Squadron-Leader in command of 610 engaged in sweeps over France as high-level bomber escort, receiving congratulations from the 12 Group Commander, Trafford Leigh-Mallory. On one sweep the squadron was attacked near Le Touquet, leading to a dogfight with three 109's, without incurring any damage. Later over Calais he had a fierce engagement with two 109's, scoring hits and probably destroying one. After being fired on by ground forces and being hit, he opened the cockpit canopy ready to eject, and his goggles were shot off, to be found later on the cockpit floor. The Spitfire limped back to Friston where he crash landed with an oil-drenched windscreen and glycol streaming from the aircraft, Ken sustaining personal shrapnel wounds to legs and forearm.

On 22 July, 1941, they were detailed to search for the German pocket-battleship 'Scharnhorst' reported making a spectacular sea dash in the coastal waters off Norway. On the 24 July 1941, he was awarded the Distinguished Flying Cross.

The 9 August was not a good day for the Tangmere Wing. Buck Casson and Douglas Bader were both reported missing after a sweep over France. The tail of Bader's Spitfire was severed aft of the cockpit during an encounter with enemy 109's, and he had to bale out hurriedly, leaving behind one of his legs which had jammed in the controls, but he landed safely near St. Omer. Buck Casson was also safe, but a prisoner-of-war. On 19 August the wing escorted a force of Blenheims on a normal day bombing mission over France, the leading Blenheim flying low over the St. Omer area to drop an artificial leg for Bader before going on to the target. Although the Germans had offered safe conduct for a small aircraft to fly across France to drop the leg, the Air Ministry declined the offer. At this time Bader's personal score was 22½ confirmed victories.

Kens activities from hereon were mainly instructional and administrative as he progressed in seniority. His log book records:-

November 4 to June 5 1942	'To 12 Group HQ. Watnall as Group Tactics Officer instructing new controllers in Sector 'Wing practice with 610, 616 and 411 Squadrons.' 'In charge of Army - Air Force Co-operations (air to ground attacks'). Visited various stations for briefings!
6 June to 27 May 1943	To Hutton Cranswick to command and re-group 610 Squadron. Visited various stations. Army exercises on moors.
12 January	Gas course at RAF Gas School, Rollston Camp.
19 April 1943	Fighter night over Hull.
27 May 1943	To Northolt HQ Fighter Command, instructing in Operations tactics (Army/Air Force co-operation.)
8 December to 14 February 1944	To RAF Staff College, Gerrards Cross. Hard work - but interesting. Had the good fortune to be there when Montgomery, Eisenhower and Lord Mountbatten visited.
6 March 1944	To Cranwell for Advanced Officer Training (Deputy Commandant). Lectured on operational flying, and day/night fighter tactics. Visited Manston.
January 1945	Posted to Tactical Air Force at Colerne
2 May1945	Airlifted from Croydon to Le Bo.
6 May 1945	Airlifted to Frankfürt.
15 July 1945	Proceeded to Detnold in Fordson Utility. Provided with one day's K rations. Set up base for Air Staff SHAEF Continuity Party. Witnessed German surrender in Stadium. They were well behaved.
31 July 1945	Advance party to Berlin. In charge of party preparing for main group arriving from Detmold on 3.8.45. Landed at Gatow airport. Visited Chancellery, Hitlers Bunker — strewn with photographs and completely stripped by Russians. Visited baby farms — scores of cots. Visited Belsen.

March 1946	Back to Doncaster — demobbed. (Rank: Group Captain)
June 1946	616 reformed at Finningley (5.6.46) Given permanent commission with the rank of Squadron Leader.
1947	Visited local stations for practices.
1950	Retired 8.12.50. Received OBE. After $4^1/2$ years, 616 Squadron became the best all-round squadron in the Royal Auxiliary Air Force - winning the Esher Trophy.

The famous trio: AVM Johnnie Johnson, Group Captain Douglas Bader and Group Captain Ken Holden
(Photo: Courtesy Wakefield Express, Wakefield)

14

'NICK'

A summarised narrative on the life and career of James Brindley Nicolson VC. DFC.

The award of Britains highest military decoration to 'Nick' — as his wife Muriel and close friends knew him — came as a result of an action against the enemy in late summer of 1940, the official citation and recommendation from his Commanding Officer, endorsed by Air Officer Commanding-in-Chief, Lord Dowding, GCVO., KCB., CMB., is described as follows:-

"During an engagement with the enemy on 16 August 1940, in the neighbourhood of Southampton, Flight Lieutenant Nicolson's aircraft was hit in the cockpit by four cannon shells; he was wounded by two of the shells, whilst another set fire to the gravity tank. Just as he was going to bale out he sighted another enemy aircraft — an Me. 110 — which he attacked and shot down into the sea. As a result of staying in his burning aircraft, Flight Lieutenant Nicolson sustained serious burns to his hands, face, neck and legs.

Flight Lieutenant Nicolson has always shown great enthusiasm for air fighting, and this incident shows that he possesses courage and determination of a high order".

This award of the Victoria Cross transpired to be the only one granted to a pilot of Fighter Command in World War 2.

Nick was a Londoner by birth but raised in the family home at Shoreham-by-Sea. He joined the Royal Air Force in October, 1936, as a pupil pilot on a four-year short service commission, his rating on eventual passing out was 'above average'.

In August 1937 he was transferred to No. 72 Squadron at RAF Church Fenton flying Gloster Gladiators, an aircraft for which he developed a great affinity.

It was at Church Fenton that he was to meet his future wife, Muriel, at a Mess dance.

After completing the course of advanced training he was transferred to No. 249 Squadron, commanded by Squadron Leader John Grandy (later Sir John Grandy, Marshal of the RAF). The squadron was currently being formed and the training was intense in order to meet Air Ministry schedule. In July, 1940, the squadron was pronounced operational (please refer to Chapter 4) and moved south to Boscombe

Down to supplement the hard-pressed squadrons of Fighter Command engaged in the 'Battle' against the Luftwaffe.

Following the action of the 16 August which led to the award of the Victoria Cross, Nick was again transferred north to Church Fenton for a period of rehabilitation and recuperation, since his badly burned hands had not regained normal mobility. His duties at this time were as Chief Ground Instructor with the newly formed No. 54 Operational Training Unit (please see Chapter 8). However, Nick was never happy tied to a desk job, and persistently asked for transfer to active duty. In September his request was granted. He moved to Hibaldstowe on experimental night fighter duties. However the unit was disbanded in January 1943.

Shortly after he was promoted to the rank of Acting Wing Commander and posted overseas to Headquarters 293 Wing, Alipore, India, where he assumed command in May, in an administrative role.

A further transfer in August to command No. 27 Squadron — known as the 'Flying Elephants' — was more to his liking and temperament. The squadron task was to harass and destroy Japanese supply lines and communications by flying sorties at zero feet over the dense jungles of Burma.

In recognition of his fine personal example, courage and enterprise in leading his squadron on the most hazardous sweeps against the Japanese, he was awarded the Distinguished Flying Cross.

Whilst serving later with the 3rd, Tactical Air Force based at Comilla, Bengal, he took part, as observer, in a bombing mission on enemy targets. Flying in Liberator KH.310, the aircraft developed engine trouble and crashed in the Bay of Bengal. Only two survivors were found by the rescue Catalina; tragically, Nick and the remainder of the crew were lost.

(Sic transit gloria mundi)

Editors footnote:
A definitive, in-depth profile depicting the life and career of this gallant young man will be published towards the end of 1990.

Wing Commander James Nicolson VC. DFC.
(Photo: Courtesy of Mrs. Muriel Nicolson)

15

EPILOGUE

A tribute to the fallen airmen of Royal Air Force - Church Fenton
by
the Right Reverend C.C. Barker
The Bishop of Selby

The noblest moments of our history are invariably associated with ordinary people who have risen above their normal performance, to do great deeds. They have not been super-men, of a different order from the rest of us, for if they were they could neither inspire nor challenge us by their example. Their impact and influence upon history has been made by the fact that they have shown the heights to which human valour, self-sacrifice and devotion to duty can rise.

The story told in this book is of a great time in our national history, and of some of the people who helped to make it so. It is a very human story, which will be read with interest and pride. I have no doubt that it will be an inspiration to all who read

Group Captain W.G. Abel at the Airmens Cemetery — Remembrance Sunday, 1972
(Photo: Courtesy RAF Church Fenton)

For 60 years, Airwork Limited has provided technical support and training services for the British and Overseas governments, international organisations, scientific bodies, private industry, civil aviation and military authorities.
Airwork employs nearly 3,000 people of various occupations operating, servicing and supporting a wide range of aeronautical and electronic equipment from basic training aircraft to supersonic fighters and from land line telephones to satellite tracking stations. Airwork establishes and manages flying training institutes and civil airports and undertakes specialist tasks at military airfields.
We offer employment for ex-service officers and technicians, who wish to continue their profession after termination of their service career.
Apply for further details to:
Mr. D. Milne, Airwork Limited, FREEPOST, Hurn, Christchurch, Dorset BH23 6BR.
Airwork Limited ~the most experienced organisation of its kind in the world
Airwork Limited
1983

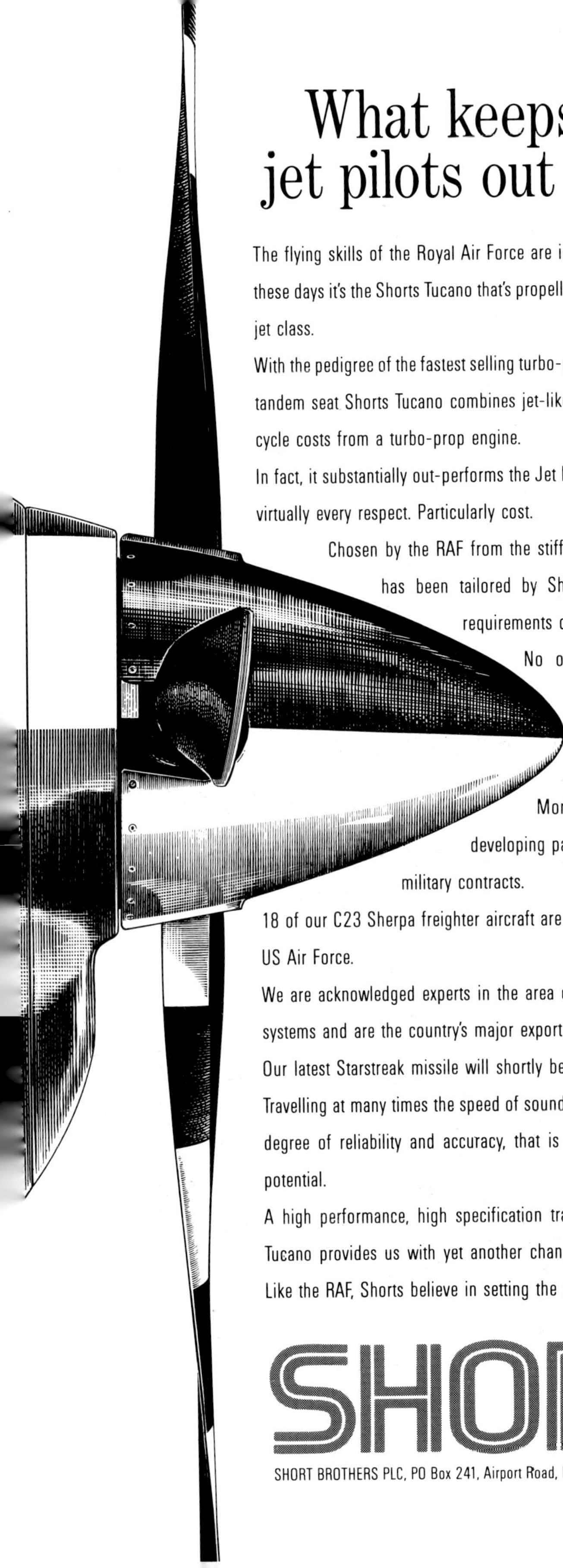

What keeps RAF jet pilots out in front?

The flying skills of the Royal Air Force are internationally admired. And these days it's the Shorts Tucano that's propelling our young pilots into the jet class.

With the pedigree of the fastest selling turbo-prop trainer in the world, the tandem seat Shorts Tucano combines jet-like performance with low life-cycle costs from a turbo-prop engine.

In fact, it substantially out-performs the Jet Provost trainer it replaces in virtually every respect. Particularly cost.

Chosen by the RAF from the stiffest competition, the Tucano has been tailored by Shorts to meet the rigorous requirements of the Ministry of Defence.

No other Air Force laid down such tough specifications.

The Shorts Tucano meets them all.

More and more, Shorts are developing particular strength in fulfilling military contracts.

18 of our C23 Sherpa freighter aircraft are currently in service with the US Air Force.

We are acknowledged experts in the area of close air defence weapon systems and are the country's major exporter of missiles.

Our latest Starstreak missile will shortly be added to Britain's defences. Travelling at many times the speed of sound, Starstreak has an awesome degree of reliability and accuracy, that is matched only by its selling potential.

A high performance, high specification trainer aircraft like the Shorts Tucano provides us with yet another chance to show what we can do.

Like the RAF, Shorts believe in setting the pace.

SHORTS

SHORT BROTHERS PLC, PO Box 241, Airport Road, Belfast BT3 9DZ, Northern Ireland.

HOW THE RAF PILOTED THE GAS CONNECTION

Thanks to the RAF, Church Fenton has been provided with its first Natural Gas supply.

More than 120 households in the village are already enjoying the advantages which come with gas connection.

British Gas and RAF Church Fenton joined forces recently to discuss the most viable means of supplying gas services to the Air Base.

And as a result a scheme was chosen whereby the entire village – as well as the Base – could be connected.

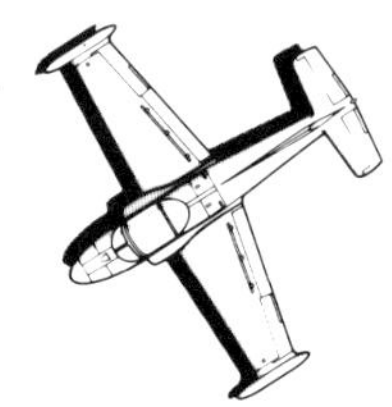

Which means that both parties are now reaping the benefits of highly economical and convenient heating, cooking and hot water.

To find out how little it costs to join up call in at your local British Gas Showroom.

British Gas
North Eastern
ENERGY IS OUR BUSINESS